Reframing Rights

Basic Bioethics

Arthur Caplan, editor

A complete list of the books in the Basic Bioethics series appears at the back of this book.

For information about special quantity discounts, please email special_sales@mitpress.mit .edu.

This book was set in Sabon by Graphic Composition, Inc. Printed and bound in the United States of America.

Library of Congress Cataloging-in-Publication Data

Reframing rights : bioconstitutionalism in the genetic age / edited by Sheila Jasanoff.
    p.   cm. — (Basic bioethics)
Includes bibliographical references and index.
ISBN 978-0-262-01595-0 (hardcover : alk. paper) — ISBN 978-0-262-51627-3 (pbk. : alk. paper)
1. Genetics. 2. Bioethics. 3. Human genetics—Social aspects. 4. Genetic engineering—Political aspects. I. Jasanoff, Sheila.
QH438.7.R438   2011
174'.957—dc22
                                                            2010053497

10 9 8 7 6 5 4 3 2 1

Reframing Rights

Bioconstitutionalism in the Genetic Age

Edited by Sheila Jasanoff

The MIT Press
Cambridge, Massachusetts
London, England

Contents

Series Foreword

Glenn McGee and I developed the Basic Bioethics series and collaborated as series coeditors from 1998 to 2008. In fall 2008 and spring 2009 the series was reconstituted, with a new editorial board, under my sole editorship. I am pleased to present the twenty-ninth book in the series.

The Basic Bioethics series makes innovative works in bioethics available to a broad audience and introduces seminal scholarly manuscripts, state-of-the-art reference works, and textbooks. Topics engaged include the philosophy of medicine, advancing genetics and biotechnology, end-of-life care, health and social policy, and the empirical study of biomedical life. Interdisciplinary work is encouraged.

Arthur Caplan

Basic Bioethics Series Editorial Board

Joseph J. Fins
Rosamond Rhodes
Nadia N. Sawicki
Jan Helge Solbakk

Acknowledgments

An edited volume is one of the most self-denying projects that academics can undertake. It disciplines space and time in ways that can try the most patient souls. It requires contributors to direct their energies toward making the whole stronger than the sum of the parts. Arguments may need to be refined and trains of thought jettisoned if they do not advance the communal project; aspects of personal voice and style may have to be scrapped if they are inconsistent with the overall tone. None of this can be achieved unless the contributors feel a genuine esprit de corps. When a good collaboration does happen, as in the case of this volume, the results are exhilarating because ideas take root and flourish that very likely could not have sprung from a single mind.

In editing this book, I had the benefit of collaborating with an exceptionally talented and dedicated group of colleagues and friends, each of whom decided that seeing the project through to completion was rewarding enough to merit considerable self-sacrifice. All but Ingrid Metzler and Brian Wynne were students or fellows at Harvard University during the period when they conceptualized their respective chapters. Most were affiliated with the Program on Science, Technology, and Society (STS) at the John F. Kennedy School of Government. Although they passed through Harvard at different times, with little overlap, they repeatedly came together in Cambridge and elsewhere to discuss their papers and identify common threads. The ideas that hold the book together, including the enveloping theoretical concept of bioconstitutionalism, emerged directly or indirectly from many bilateral and multilateral conversations within the group. The authors cheerfully complied with my multiple editorial requests, put up with unforeseen delays, and responded in good humor to the most pointed reviewer comments. A panel on bioconstitutionalism at the 2010 annual meeting of the Society for Social Studies of Science in Tokyo, attended by several of the authors, served as a capstone for the project. I owe all of the contributors a tremendous debt. Without their unflagging support and encouragement, this collection could never have come into being.

As with most of my research and writing, support from the National Science Foundation (NSF) played a crucially important role. The book grows to varying degrees from two NSF grants: Reframing Rights: Constitutional Implications of Technological Change, NSF Award No. SES-9906834; and Sociotechnical Imaginaries and Science and Technology Policy: A Cross-National Comparison, NSF Award No. SES-0724133. The Reframing Rights grant helped train seven of the authors during their stints as STS Fellows at Harvard. Apart from furthering their research in STS and law, the grant importantly advanced the fellows' careers, leading in almost all cases to appointments at major research universities. My coauthors and I gratefully acknowledge NSF's generous support.

Several individuals deserve to be thanked by name. At NSF, Rachelle Hollander was the program director who oversaw the Reframing Rights award; in this, as in so many instances, her ability to see beyond known horizons enabled a group of STS scholars to define a new area of research. My former Kennedy School colleague Frederick Schauer was coprincipal investigator with me on the Reframing Rights grant. Though our paths diverged when he left Harvard, his collegiality was key to our receiving and successfully managing the grant. Ben Hurlbut, though not a formal contributor to this volume, was a constant interlocutor with respect to its key ideas over several years. His 2010 Harvard dissertation, *Experiments in Democracy: the Science, Politics and Ethics of Human Embryo Research in the United States, 1978–2007,* is as much a work in the framework of bioconstitutionalism as any of the chapters in this book. Ori Aronson offered indispensable pointers on current debates in constitutional law that are deeply relevant to issues raised in this book. My assistants Lauren Schiff and Shana Rabinowich provided invaluable help in preparing the final manuscript. Clay Morgan at the MIT Press was as always a thoughtful, encouraging, and supportive editorial presence. And three anonymous reviewers, sharp but helpful, contributed greatly to the tightness and clarity of the final manuscript. The authors of course accept responsibility for any remaining errors and omissions.

1

Introduction: Rewriting Life, Reframing Rights

Sheila Jasanoff

We hold these truths to be self-evident, that all men are created equal, that they are endowed by their Creator with certain unalienable Rights, that among these are Life, Liberty and the pursuit of Happiness.

—*U.S. Declaration of Independence*, July 4, 1776

The Palimpsest of Life

Two encyclopedic bodies of writing—one social, the other scientific—define the meaning of life in our era. Encompassing, respectively, law and biology, these intertwined, mutually supporting, indeed coproducing textual projects frame the possibilities, limits, rights, and responsibilities of being alive—most especially for the species we call human.[1]

Law from ancient times has been a matter of wording. "In the beginning was the word": first God's word and then our own secular texts, collectively agreed on. Legal writing makes visible the rules of action and behavior that human societies accept as controlling; it is the legibility of the law, in short, that makes it intelligible and thereby enforceable throughout a society that submits to the constraints of civilization. From the code of Hammurabi and the Ten Commandments of Moses to the laws of Manu and the edicts of the Emperor Ashoka, from Magna Carta to the French Declaration of the Rights of Man and the American Declaration of Independence, emerging communities have signaled their solidarity by promulgating new, legally binding texts. Since the earliest recorded histories, writing or inscribing the law has been a political enterprise, and rulers have taken great pains to ensure that people will read the law and learn how to comply with its authoritative mandates.

Textuality in the life sciences is of much more recent vintage. Compared with five millennia of law writing, the association of biology with written texts occupies a blink in time, but its implications for human rights and entitlements have been no less momentous. The textual phase of the modern biological sciences began in 1953, with the discovery of the structure of DNA (Kay 2000; Keller 2000; Watson and Crick 1953). This was the revolutionary moment when it became possible to represent the basic matter of life with permutations and combinations of just four letters of the Western alphabet: A, T, C, and G. Those letters, of course, stand for structures a great deal more complex. Adenine (A), thymine (T), cytosine (C), and guanine (G) are the names of four chemical compounds, called *bases*, which bond in pairs along the sugar-phosphate backbone of the now-familiar DNA double helix. Separated, each strand becomes a template for generating its precisely ordered partner: unwound and repartnered in hospitable biochemical environments, a single segment of DNA gets remade as two identical helices, thereby supplying the mechanism of replication that had puzzled students of heredity for decades. The sequence of bases, the length of the strands, the relative stability or instability of bonds, and many other factors of developmental and environmental biology affect the transformation of DNA into the wild profusion of living organisms known in nature. Yet the elegant simplicity of the four letters, capable in principle of generating untold varieties of new life, enables a discourse of information and rule following that makes biological heritability converge in salient respects with the normative functions of the law (Kay 2000, 1993; Kevles and Hood 1992; Lewontin 1993).

Advances in biological knowledge and technique may in the future dilute the impact of the "book" of DNA. Knowing life may some day become more a matter of tinkering with it than reading it. Synthetic biology—a focus of rising scientific, political, and economic attention since the turn of the century—promises gains by designing and engineering life rather than by decoding its informational content. Through the construction of novel biological parts and the instrumental use of natural biological materials, synthetic biology redirects the understanding of life into distinctly material channels. Both scientists and policymakers see enormous potential in the turn to engineering (Keasling 2005; Specter 2009), and the entrepreneur J. Craig Venter's early experiments to create life's smallest units aroused enthusiasm while also raising eyebrows (Hotz 2010). Synthetic biology's building projects, however, derive their blueprints from the world of imagined configurations opened up by the texts of DNA.

This collection of essays is located in the overlapping spaces created by a half-century of rewriting life in genetic sciences and technologies and the centuries-old texts of law that represent one of the most durable monuments of human culture. It is impossible today to understand the ethical debates surrounding the life sciences without looking deeply into the evolving relationships among biology, its technological applications, and the law. This is not, as is sometimes thought, a one-way relationship, with science always leading the law. Even when biological advances seem most surely to be putting new issues on the agenda—as, for example, in conflicts over the moral status of human embryonic stem cells or the ownership of novel biological materials—powerful legal norms lie barely concealed beneath the surface, conditioning the very terms in which those debates are formulated (Jasanoff 2001). The constant, mutually constitutive interplay of biological and legal conceptions of life, the former focusing on life's definition and the latter on its entitlements, is a fundamental feature of scientific and technological societies; it exemplifies the coproduction of *is* and *ought* in modern times (Jasanoff 2004).

The frictions and ambiguities recorded in the palimpsests of law and science become concrete when biological knowledge is translated into material form—that is, into tangible, working components of biotechnological systems. Throughout this collection, we see questions raised by new entities, objects, techniques, and practices that embody genetic understandings of life, but whose legal and social meanings are far from clear at the moment when scientific work first conceives of them or, through material transformations, brings them into being. Conversely, we encounter quite different articulations of what societies value about "Life, Liberty and the pursuit of Happiness" as diverse legal institutions and cultures struggle to make sense of biological claims, materials, and practices that destabilize the law's well-made conceptual categories.

This book argues that periods of significant change in the life sciences and technologies should be seen as constitutional or, more precisely, *bio*constitutional in their consequences. Revolutions in our understanding of what life is burrow so deep into the foundations of our social and political structures that they necessitate, in effect, a rethinking of law at a constitutional level. At these moments, the most basic relations between states and citizens are reframed through changes in the law (Jasanoff 1987, 2001, 2003, 2005). Reframing begins with redefining human life but segues into redefining the obligations of the state in relation to lives in its care. Just as the translation of DNA to RNA to protein has been called the central dogma of molecular biology, so safeguarding the lives of citizens can be regarded as the central dogma of the constitutionally regulated state. Put differently, the first duty of any state committed to the rule of law is to take responsibility for its people's lives; indeed, the legal philosopher H. L. A. Hart (1961) defined the "minimum content of natural law" as that content which assures the survival of the society that the law seeks to regulate. Radical shifts in the biological representation of life thus necessarily entail far-reaching reorderings in our imagination of the state's life-preserving and life-enhancing functions—in effect, a repositioning of human bodies and selves in relation to the state's legal, political, and moral apparatus.

Such transformations do not happen all at once, or coherently. The ongoing work of constitution making during a scientific revolution is patchy, slow, and unpredictable, but we can piece together its emerging outlines and principles by looking closely at specific locations in which law, the life sciences, and biotechnologies have entered into conflict or conversation. These range from highly focused controversies about the rights and duties of living entities (e.g., what are the legal rights of stem cells; what new rights attach to DNA typing?) to abstract issues of democratic legitimacy (e.g., if the definition of life is itself in flux, how should the state construe its responsibilities to preserve life; and how should it deal with moral uncertainty and conflict?). Through snapshots of legal developments in North America, Europe, and India, this book seeks to capture the dynamics of the contemporary bioconstitutional moment as it is unfolding in real time and globalized space.

Our use of the label bioconstitutionalism to describe these aggregated movements was itself the result of incremental observation and analysis. We did not begin with this term in mind but concluded, over several years of exchange and mutual provocation, that it captures much of what is salient in today's life-law interactions. These essays, the result of the authors' thinking together, are grounded in the field of science and technology studies (STS), but with wide-ranging implications for bioethics, law, and political theory. STS research is often criticized as insufficiently theoretical and deaf to normative problems (Jasanoff 2004). We show to the contrary that investigations of biological and legal change are inseparably linked to fundamental questions about justice and social order. The book as a whole offers a programmatic way of looking at the nexus of law and science, taking on board the influence of science and technology on basic categories of legal thought, and vice versa. Individual chapters can be read as stand-alone pieces that exemplify this broader project; they also illustrate varied approaches to studying the interoperability, as it were, of biological knowledge and constitutional norms.

Substantively, the authors investigate cross-cutting transformations in law and biotechnology that are altering how human societies think about what it means to live and to be human, and what rights and values attach to human-ness or to living. We consider how the law responds when new biological constructs cross conceptual boundaries that the law previously took to be natural—for instance, boundaries between life and nonlife, human and nonhuman, individuals and collectives, and predictable and nonpredictable risks. As these lines blur, we see corresponding instability in legal thinking about the rights, duties, entitlements, and needs of living entities. The chapters display the struggles and realignments involved in attempts to restore epistemic and normative order under uncertain circumstances.

Theoretically, the book opens new ground on four interrelated fronts: in legal scholarship, science and technology studies, comparative politics, and bioethics. For constitutional scholars, our approach offers new ways of reading the relationship between science, technology, and the rule of law, with greater sensitivity to the value-ladenness of novel biological entities and practices. For academic students of science and technology, we map the major intersections between life sciences, biotechnologies, and the law, interpreting these through the lens of coproduction and teasing out their normative implications. For theorists of legal and political culture, the essays illuminate distinctive features of national (and in some cases supranational) politics, public reason, and decision making that inflect the legal treatment of life in the post-DNA era. For the still young field of bioethics, these essays open wider horizons by showing how ethical principles are not neutrally applied to biotechnology's cornucopia of novel entities and practices, but are instead reformulated and redefined through ongoing processes of ontological clarification.

We first offer a brief history of the intersections between the texts of life and the texts of law that have given rise to questions about rights, entailing the reframing moves flagged in the title. Next, we situate our distinctive view of constitutionalism against a backdrop of legal thought, juxtaposing constitutionalism in its conventional senses with the *bio*constitutionalism that the authors explore, often in cross-cultural perspective. A consistent theme is the impossibility of any form of deterministic analysis of relations between law, science, and technology: changes in biological understanding do not ineluctably shape the law; nor do law and ethics prescribe inviolable limits to scientific and technological advances. Chapter summaries round out this introduction.

Life under Law

In a monumental corpus spanning the histories of madness, sexuality, and other forms of socially controlled behavior, Michel Foucault developed the influential concepts of biopower and biopolitics. Observing that the contemporary state exercises power not by commanding the deaths of dissidents but by regulating the bodies and lives of consenting subjects, Foucault spoke of "an explosion of numerous and diverse techniques for achieving the subjugations of bodies and the control of populations" (1998, 40). The harsh characterization of human subjects as nothing more than subjugated bodies, absorbed into the amorphous mass of the population, softened in Foucault's later writing. His followers too see the governance of lives as more of a two-way street—exposing subjects to state classification and control to be sure, but also creating scope for new forms of voluntary association facilitated by shared biological characteristics. Using terms such as biosociality (Rabinow 1992) and biological citizenship (Petryna 2002; Rose 2006), social theorists of the genetic era have sought to capture, and to some degree celebrate, the opening up of agency from below. Heredity, they argue, no longer equals destiny as in the bad old days of state-sponsored eugenics (Kevles 1985; Kevles and Hood 1992). Instead, genetic texts and instruments offer individuals a chance to retranscribe their own solidarities and destinies with newly acquired knowledge and technologies.

All such possibilities for self-fashioning play out on terrain already occupied by law. Concerns about the need to regulate the disruptive potential of biological manipulation were apparent almost from the moment when genetic engineering became feasible in the 1970s. There was a recognition that recombinant DNA techniques cross species and organismic boundaries, and that these crossings and hybridizations may pose significant legal problems. With time, it became clear that genetic technologies transgress more than one kind of boundary, with implications for many different domains of law. Thus, early worries focused mainly on the safety of genetically modified organisms and translated largely into matters of administrative and regulatory concern. Subsequently, several additional boundaries gained prominence: between life and nonlife, with associated issues of commodification, ownership, and property law; between human and nonhuman, entailing questions of the moral status of biological constructs; between individual and group rights; and between uncertainty and predictability, implicating the custodial responsibilities of states for the societies they govern.

In the United States, where the new biotechnology was first invented, themes of biopower and biopolitics were relatively slow to emerge. Instead, a polity already attuned to nuclear and chemical hazards assimilated genetically modified organisms into familiar imaginaries of risk and regulation (Jasanoff and Kim 2009). Precisely snipping bits of DNA from one organism and transposing them into others, using enzymes as molecular "scissors," promised to endow living things with valuable new traits, but the process also raised fears that the resulting entities might escape from the contained environment of the laboratory and play havoc with human health and the environment. Scientists were among the first to raise the alarm. The February 1975 Asilomar Conference on Recombinant DNA (rDNA), convened by the Stanford molecular biologist and future Nobel laureate Paul Berg, outlined regulatory principles for governing rDNA research (Berg et al. 1975). That conference, memorialized simply as *Asilomar*, became the twentieth century's iconic example of scientific self-regulation, the antidote to Los Alamos. It was, however, an achievement based on a narrow framing of biosafety (Jasanoff 2005). Participants focused mainly on risks to lab workers and surrounding communities, and on producers' liability for possible harms. The principles they drafted were aimed primarily at containing the spread of novel organisms through biological and physical means. Absent from the agenda at this dawn of regulation were challenging questions about how to classify the entities created by gene splicing, how to manage the impacts of industrial biotechnology on agriculture and species diversity, and who should set limits on the purposes, ambitions, and scope of genetic interventions (Gottweis 1998; Jasanoff 2005; Wright 1994).

Legal imaginations and horizons widened significantly in the 1990s. The launch of the federally funded Human Genome Project (HGP) in 1990 drew renewed attention to the informational content of the genetic code and stirred up debates about privacy, stigmatization of genetically marked persons or populations, and the misuse of genetic data by employers, insurers, and law enforcement agencies. Scholars and journalists wondered aloud whether ethical principles and legal rules would evolve in time to keep the new scientific discoveries in check. Committees were charged with evaluating the ethical implications of research funded by the National Institutes of Health (NIH), the chief grant-making arm of the U.S. government for biomedical research. James D. Watson, codiscoverer of the structure of DNA and first director of the HGP, responded to these swirling anxieties with an off-the-cuff promise to commit 3 percent (later 5 percent) of project funds to examining the Ethical, Legal, and Social Implications of genome research. That endeavor, institutionalized as the ELSI Program, served as a model for later national and international initiatives.

The middle of the decade ratcheted up public concern. In 1997, researchers at the Roslin Institute in Edinburgh announced that they had cloned a sheep from cells taken from the mammary glad of an adult ewe, the first time that a genetically identical, mammalian offspring had been asexually generated from a parent animal. Dolly the sheep, jestingly named for the voluptuous country music singer Dolly Parton, lit a fuse of ethical anxiety laid decades ago. The scientific term *clone* carried uncomfortable baggage outside the laboratory. Dystopic fantasy films such as *Boys from Brazil*, about the Nazi doctor Josef Mengele's plot to rule the world with an army of Hitler clones, and *Clonus*, in which human clones were bred to supply spare organs for the wealthy, had circulated in popular culture since the 1970s, veneering scientific potential with moral dread. The possibility of creating identical copies of nonhuman mammals morphed easily into nightmare visions of industrially manufactured, intellectually subjugated, enslaved human beings, like the subhuman populations rolling off the assembly lines in Aldous Huxley's 1932 novel *Brave New World*.

Succeeding years saw an explosion of public doubt ranging far beyond Asilomar's fixation on the containment of rDNA risks. As biomedical practices raced to convert therapeutic visions into reality, it became clear that the textuality of the genetic code was not simply a metaphor but also an enabling instrument. In effect, the code had rendered life programmable, or subject to design: it seemed increasingly probable that humans and other living things could be selected for, or actively engineered with, designer traits to make them longer-lived, more athletic, endowed with particular forms of beauty, or otherwise attractive to designers' imaginations. Suddenly, how far humans should go in enhancing their own genetic constitution arrived at the forefront of academic reflection (Buchanan et al. 2000; Fukuyama 2002; Kitcher 1996; Sandel 2007). While scholars debated whether some essence of human-ness should be left untouched, pragmatic minds turned toward correcting perceived genetic errors with techniques enabled by genetic research. An early, controversial instance was the creation of "savior siblings," selected as embryos through a technique called *preimplantation genetic diagnosis*, to serve as tissue donors for existing children with incurable genetic illnesses. Permitted in some countries[2] and prohibited in others, the practice of testing and selecting artificially created human embryos to treat their own close kin underscored both the instrumental potential of the genetic code and the lack of legal and ethical clarity surrounding its possible applications.

That no rules were in place to regulate such interventions only confirmed for many the reality of the law lag: the claim that scientific and technological innovation inevitably proceeds at a more rapid clip than legal rule making, so that the law is doomed to lag behind the frontiers of science and technology (Jasanoff 2008). That view, however, misconceives the immanence of the law. True, the arrival of novel entities or practices often requires a fine-tuned specification of existing principles to deal with new contingencies. But law is always already present as a conceptual and cultural resource, governing responsible human behavior and conditioning the terms in which people imagine the normative organization of their worlds. We return to this point following a brief review of constitutional thought in relation to biological change.

Bioconstitutionalism: Rethinking Ontologies and Rights

From the earliest days of genetically based biomedicine, legal and policy analysts saw possible contradictions between constitutional guarantees and advances in biotechnology. Prospects of genetic testing and gene therapy fed worries about information privacy and discrimination through the creation of a genetically stigmatized underclass with reduced access to employment, health care, insurance, and other social goods (Silver 1997). Privacy and due process were debated in connection with the widespread adoption of DNA typing as a forensic technology (Kevles and Hood 1992); those questions intensified as states and private institutions went about establishing "biobanks" as repositories of personal genetic information (Häyry et al. 2007; Hindmarsh and Prainsack 2010). Questions about the limits of inquiry arose in connection with germ-line gene therapy, xenotransplantation, and the creation and patenting of transgenic animals. Dolly's birth spurred renewed reflection on the implications of mammalian cloning for human dignity, reproductive freedom, the right of governments to ban or restrict scientific inquiry, and conflicts between science and religion.

Unifying this first generation of constitutional thought was an underlying certainty, or taken-for-grantedness, about the nature and meaning of rights. Constitutional rights are typically seen as among the most stable elements of national legal systems: to be held as far as possible sacred, and to be defended against erosion by vigilant lawmakers or watchdog organizations such as the American Civil Liberties Union (ACLU). It is accepted in broad terms that we *know* what privacy is, what it means to do research without constraints, or when a search or seizure is too intrusive and unwarranted. The challenge is to discern when rights are under stress, including from new technologies, and must be reasserted. Such certainty about the nature of rights, however, depends on definitions of the nature and needs of the human subject that are typically neither questioned nor reexamined. Instead, classical constitutional thinking operates with a tacit understanding that human-ness is held constant by nature (biology), and that the law needs to respond only when those highly valued, enduring, and natural human entitlements are threatened by technological intrusions.

Such conservatism comes at a high price. To begin with, posing constitutional questions within an impacts framework feeds the perception of the law lag, because the law seems continually to fall behind in its efforts to define, preserve, and protect the rights that constitutions guarantee. And decoupling the talk of rights from the actualities of scientific research and development limits the scope of legal and ethical analysis. Scant attention is paid, for example, to the law's treatment of new biological entities and their incorporation into regimes of rights (Stone 1974)—for example, in disputes about the moral status of stem cells, the patenting of novel life forms, the ethics of producing human-animal chimeras, or the classification of transgenic species. More generally, analytic weaknesses arise from attempts in constitutional jurisprudence to make do with notions of human identity, liberty, property, and nature that predate even the industrial revolution, let alone contemporary developments in biological, informational, and environmental sciences and technologies (Schauer 1998). Emergent rights vanish from the periscope of constitutional analysis. The assumption that rights have remained the same while the world has changed around them imparts a kind of rigidity to constitutional thought and contributes to the perception that the law is unduly resistant to change.

How then should legal scholarship develop a more supple framework for addressing the constitutional implications of epochal changes in science and technology? This book lays out several theoretical and methodological avenues. To begin with, we broaden the notion of *constitution* to include the full range of sites and processes in which individuals work out their biopolitical relationships with the institutions that regulate them. This expansive frame—we may think of it as constitutionalism with a small "c"—reaches well beyond the judicial interpretation of formal legal documents such as the U.S. Constitution. It extends the notion of a "legal text" to include not only written rules and opinions, but also the institutional practices that make up a constitutional order. It takes account of science's role in producing what the legal scholar Bruce Ackerman (1983) calls *constitutional moments*: moments of radical restructuring in state-society relations that may or may not be formally ratified through constitutional amendments.[3] It also accommodates disparities among the world's written and unwritten constitutions, which vary greatly in their understandings of the human as a legal and political subject, and hence in their elaboration of human rights. Under the rubric of constitutionalism in this wider meaning, we explore the kaleidoscopic ways in which definitions of individual and collective rights both influence and are transformed by changes in the biological status of the human (Jasanoff 2004, 2008).

Our understanding of constitutionalism with a small "c" underscores some parallel theoretical preoccupations in law and STS that this book helps identify, though fuller exploration of those commonalities lies beyond our present purposes. Just as STS research has located science in mundane activities—atheoretical, habitual, done by technicians and instruments—so some progressive legal thinkers have sought to democratize constitutional thought, finding it in the actions and resistances of the "the people themselves" instead of only in principled decisions promulgated by supreme judicial authorities (Kramer 2004). Just as STS analysts have deliberated on the appropriate balance of power between expert and lay understandings of knowledge and norms, so legal scholars have been wrestling with the proper relationship between professional legal reason and popular legal thought or "democratic constitutionalism" (Post and Siegel 2007; see also Aronson 2009–2010; Harding 2006; Waldron 1999). At a deeper level, both STS and legal theory are perplexed by shared questions about truth and finality that resonate throughout this volume: what counts as right, in knowledge and action, and who has the right to declare it so?

A second definitive step that this book takes is to move away from the framework of technological determinism that the law often adopts as its own. This theory attributes causal force to material technologies, so that changes in society are seen as results of ongoing changes in technology (Smith and Marx 1994). Technological determinism underlies many familiar assertions about modernity, for example, that automobiles dispersed people into the suburbs and fragmented families, television dissolved communal solidarities, and social networking technologies such as Facebook and Twitter increased personal and political freedom. Deterministic ideas are at play whenever the law is depicted as trying to bridge gaps and lags created by advances in science and technology. Most important for this book, determinism surfaces whenever technology is seen as infringing on and eroding well-established rights. Constitutional rights tend, at such moments, to be construed as passive guarantees instead of as active conceptual agents shaping the very meanings that we attach to technological artifacts and practices. We argue that it is important to be attentive to the reciprocal moments in which legal sense-making influences biological categories—by placing entities on one side or the other of normatively meaningful divides such as natural-unnatural, living-nonliving, or human-nonhuman.

Much of contemporary bioethical concern with human rights is rooted in deterministic thinking. In the course of the genetic revolution, reductionism (sometimes labeled "geneticization") became a major worry. The specificity of the genetic code invites redefinition of the most complex biological organisms and their developmental potential in terms of that code's seemingly inexhaustible alphabet. From plants to animals to humans, genetic characterizations then become paramount: Bt corn, so labeled for its inserted insecticidal gene; the oncomouse, named for its genetically modified susceptibility to cancer; bearers of sickle cell, Huntington's disease, or breast cancer genes, known to insurers and employers for those traits above all others (Duster 1990; Kay 1993; Keller 1992; Lewontin 1993). Although such critiques display an admirable understanding of the uptake of scientific representations into society, insistence on reductionism buys into the paradigm of technological determinism. Bioethicists, some have argued, contribute to the apparent inevitability of "geneticization" by focusing exclusively on the rights of persons characterized by undesired genetic traits (Árnason and Hjörleifsson 2007).

By contrast, bioconstitutionalism, as elaborated in this volume, stresses the irreducible contingency of life-law relationships and thereby helps restore normative agency to social actors. In this respect, bioconstitutionalism complements work in critical studies of the law. Legal theory has been hugely influential in bringing to light the contingency of legal rules, illuminating the hidden normative assumptions that underpin supposedly neutral legal rules and potentially influence rule-following behavior. Modern versions of legal realism have refocused the understanding of realism away from the indeterminacy of rules toward the often-disguised substantive choices embedded in even relatively determinate rules (Fisher, Horwitz, and Reed 1993; Fried 1998; Kysar 2010). Feminist jurisprudence exposes the gender-based assumptions that support dominant, male legal understandings (Bartlett 1990). The Critical Legal Studies movement stressed the ideological contingency of legal structures that may appear inevitable and natural (Kairys 1990; Kelman 1987). And with regard to constitutional decisions, legal scholars have questioned the neutrality and validity of the "baselines" against which we consider constitutional questions, contending, for example, that the distinction between state action and private action presupposes a state-created status quo that established the domain of the *private* in the first instance (Sunstein 1993).

Despite these turns toward reflexivity, legal scholars have not by and large grappled with the ways in which legal rulemaking interacts with the life sciences and technologies to build the concept of rights. Even at its most sensitive and reflective, legal scholarship tends to accept the separation between law-work and science-work, seeing the former as normative and the latter as epistemic; similarly, the notion that technological objects may have norms built into them is not widely acknowledged in legal scholarship, though standard in STS. Rather, law and science are seen most often as distinct "cultures" that clash when they meet in disputes over rules and policies (Goldberg 1994; Schuck 1993). An innocent positivism still marks much writing about science and the law, exemplified by a stream of work criticizing judges, juries, Congress, and even expert agencies for failure to abide by the standards of good science (Breyer 1993; Foster and Huber 1997; Huber 1991). Such critiques often accompany triumphalist and historically inaccurate accounts of technological progress, which represent the law not only as lagging, but also as an awkward impediment to human betterment through science and technology.

The separatist tendency prevails even though the historical record suggests that law and science have supported each other for centuries in patterns of mutual construction, stabilization, and reinforcement (Ezrahi 1990; Porter 1995; Shapin 1994; Shapin and Schaffer 1985). There has been relatively little systematic reflection on the ways in which modes of authorization in science and the law build upon, mimic, or incorporate one another (for exceptions, see Jasanoff 2005, 2008; Smith and Wynne 1989; Wynne 1982, 1988).

Cutting against the deterministic tendencies of much legal analysis, work in science and technology studies has consistently shown that the products of technoscience not only influence but also incorporate and reaffirm social values and institutional practices (Jasanoff 1995, 2004, 2008; Jasanoff et al. 1995). Nuclear power plants, smart bombs, ozone holes, computers, genes, Dolly, and the oncomouse do not merely appear in the material world; they also manifest particular ways of imagining futures, creating social order, and ratifying moral judgments (Bijker et al. 1987; Haraway 1991, 1997; Jasanoff and Kim 2009; Latour 1988, 1993; Latour and Woolgar 1979; MacKenzie 1990). Biological artifacts engage with and reshape our perception of rights and entitlements at many levels: by redrawing the boundaries between humans and nonhumans (Callon 1986; Latour 1993), by altering fundamental notions of human identity and difference (Epstein 2007; Haraway 1997; Rabinow 1992), and by disrupting settled understandings of the state's biopolitical prerogatives (Jasanoff 2005). These insights are consistent with the views of a handful of legal scholars working on the intersections of law and technology—for example, Lawrence Lessig (1997) on the architecture of information systems, James Boyle (1992) on intellectual property, Frederick Schauer (1998) on privacy and the Internet, and most recently Douglas Kysar (2010, chapter 7), whose sophisticated assault on the objectivity of cost-benefit analysis in environmental law calls attention to the need for new sources of ethics when genetic technologies are destabilizing the basic categories for classifying living things. Systematic conversation, however, has yet to occur between these convergent strands in law and STS. This book hopes to jump-start that exchange.

Within STS, Bruno Latour offered a suggestive articulation of the concept of bioconstitutionalism, though not using that term, in his influential 1993 monograph, *We Have Never Been Modern*. There, Latour called attention to the work that human societies do to "purify" their world of hybrid networks into seemingly distinct spheres populated by pure entities of nature and culture. He termed the resulting settlement "constitutional," because it establishes the most fundamental cleavage in modern social experience: between what we make for ourselves and what is given to us by an independent nature accessible only through science. In reality, Latour argues, all of the things that define modern existence are mixed-up and hybrid, culture and nature churned up together. And yet people somehow go about unaware of this, as if categorical distinctions were simple and straightforward: "The smallest AIDS virus takes you from sex to the unconscious, then to Africa, tissue cultures, DNA and San Francisco, but the analysts, thinkers, journalists and decision-makers will slice the delicate network traced by the virus for you into tidy compartments where you will find only science, only economy, only social phenomena, only local news, only sentiment, only sex" (Latour 1993, 2). He might have added "only law." In the creation and maintenance of such neatly bounded categories, and the resulting erasure of society from nature, Latour locates the constitutional dynamics of modernity.

Powerful as these insights are, they leave many questions unanswered—questions that matter to anyone wishing to make sense of new scientific and technological goings on, let alone to shape their use or meaning. Like any universalizing theory, Latour's account of the modern is too abstractly metaphysical. It fails to account for the divergences one finds among articulations of the nature-culture boundary in different times, places, institutions, and societies. The mechanics of purification, too, remain largely unexplored in Latour's schema. One wonders how preexisting normative commitments (including those embedded in constitutional law) affect the reordering of the hybrid products of technoscience into accepted categories of natural and social. Why, for instance, did Dolly's cloning induce no frissons of disgust, whereas the use of similar techniques to clone humans, create human-animal chimeras, make glowing rabbits or blood-stained petunias, or knowingly enhance human traits produces clamor and controversy? Finally, the regime of sharp demarcations set forth in Latour's modernity seems inconsistent with the fluidity, ambiguity, and cultural heterogeneity of technoscientific constructs noted by many STS scholars (Cambrosio et al. 1990; Haraway 1991, 1997; Jasanoff 2005; Latour 1987; Mol and Law 1994; Star and Griesemer 1989).

Latour's metaphysics dwells on the separation of the natural and social orders at the highest constitutive level: the creation of modernity's sense of orderliness. In the ongoing, mundane interactions of law and the life sciences, we encounter more what I have termed the "interactional coproduction" of two already separated worlds struggling to name, define, and deal with novel ontologies that trouble their boundaries (Jasanoff 2004; also Testa, chapter 4, this volume). In the interactional register, we confront problems of normativity specific to legal regimes: what rights should humans have vis-à-vis new biological techniques that impinge on their lives; where should human agency, and the protections accorded to it, begin and end; when are humans entitled, as citizens, to participate in governing new forms of life; and who in any case should represent, or speak for, rights disrupted by advances in the life sciences and biotechnologies? These are among the concerns that we address.

The textual analysis of high court and appellate decisions—the staple of constitutional scholarship—offers at best a partial window on our concerns. Rights have to be seen as more than constructs discerned by judges trained in legal reasoning and articulated in legal language. For rights to have social meaning, they must become embedded in people's imaginations and understandings and worked out in their practical dealings with one another, with the products and processes of technoscience, and with governing institutions. A right in practice emerges not only at the moment when a court declares it, but also when people (and institutions) assume that they or others own the right and can assert it through their actions. Thus, there may be quasi-constitutional rights that no court has declared nor legislature decreed, but that are created (or constrained) through everyday practice and thought in technologically advanced societies. These include, for instance, the right to say no to particular directions in research and development, through actions that the existing legal order may see as extralegal or even illegal. In relation to the life sciences, moreover, the stream of new objects emerging from the work of laboratories and clinics plays an unavoidable part in reframing rights: such objects may extend rights in new directions, as in the protections accorded to embryos and stem cells; or constrain rights people thought they had, as when a court decides that people may not own the tissues and cells taken from their bodies (Boyle 1992); or, to the contrary, that human genes may not be patented.[4] Bringing these tacit normative presumptions to light, and illuminating the areas where disagreements lurk, will be important tasks of constitutional deliberation and cross-national comparison in coming decades.

To obtain a fuller picture of bioconstitutionalism, researchers have to dig below the level at which rights are explicitly recognized as being threatened or violated. Inquiry has to focus as well on what we view as the basic building blocks of rights: that is, on social commitments concerning what is worth protecting and why, for and against whom, through which kinds of social and institutional agency, by what means, to what extent, and through what processes. It is at this deeper level, that one may elucidate the impacts of science and technology on the very notion of rights—not only as these are formally construed by courts, but also as they are tacitly understood and worked out by scientists, lawyers, and policymakers (Jasanoff 1987, 1990); articulated in research practices (Duster 1990; Epstein 2007; Hilgartner and Brandt-Rauf 1994); hardened into material technologies, or built into professional discourses (Cambrosio et al. 1990; Jordan and Lynch 1999) and political practices (Gottweis 1998; Jasanoff 2005).

New genetic understandings and capabilities have affected notions of race, diversity, kinship, ethnic and social identity, normality, deviance, criminality, justice, and human uniqueness. Biotechnology has also created new forms of life, including plant genetic resources, embryos, stem cells, biobricks, and human-animal chimeras, along with claims of ownership and demands for state protection. Further, in an era of globalization, these developments have been caught up in changing definitions of state and of sovereignty, problematizing at one and the same time the meaning of rights and the political agents who are responsible for defining and protecting them. Shifts in the understanding of human nature, of distinctions between natural and unnatural objects, and of state prerogatives and obligations have opened up a wide array of constitutionally significant questions that the authors explore.

Emergent and Contested Rights

The theme of contingency is central to most of the contributions. In chapter 2, Alex Wellerstein takes direct aim at deterministic presumptions with a historical examination of California's notorious sterilization program in the first half of the twentieth century. California's enthusiastic embrace of sterilization is widely seen as working out a bad idea, namely, eugenics: the discredited theory that the fitness of a race can be secured through systematic barring from reproduction of its least fit members. Wellerstein shows the inadequacy of this explanation. Looking at sterilization practices in three California medical institutions, he argues that decentralized decision making, a characteristic of U.S. political culture, offers a better explanation of what happened in practice. California hospital administrators, who enjoyed enormous discretion, chose sterilization as a treatment method for idiosyncratic reasons. That incoherence also made for the technique's rapid disappearance when the state eventually centralized its administrative apparatus and removed treatment policy from local, individual control.

Subsequent essays explore the coming into being and constitutional ordering of new objects, new rights-bearing subjects, and new rights. In chapter 3, I use comparative analysis to explore why embryos and their derivatives have been treated differently in national bioethical deliberations. Sketching some of the connections between bioethics and biopolitics, I show how commitments to specific bioconstitutional arrangements influenced ethical choices in the United States, Britain, and Germany. Drawing on my observations as participant in a U.S. stem cell oversight committee, I trace how the committee's micro practices of line drawing and classification separate entities of moral concern from those that are not entitled to such deference. This process of "ontological surgery" serves as a basis for applying moral principles that appear neutral but are in reality consistent with particular preordained notions of constitutional governance in the United States. Here, as in other chapters, comparison makes visible the underlying cultures of observation and reasoning on which bioethical rules and rule applications depend.

Cross-cultural contingencies are explored again in chapter 4, Giuseppe Testa's study of the intricacies of cloning policies in the United Kingdom, Italy, and the United States. He compares three national approaches to dealing with clones derived from somatic cell nuclear transfer, bringing into relief differences in the epistemic underpinnings of legal order in the three countries. Those differences, as Testa illustrates, resulted in different conclusions about permissible and impermissible scientific research, as well as what constitutes the public good in knowledge making and who is responsible for funding it. He analyzes the place of human life in three national sociotechnical imaginaries, showing how life itself is constituted in the practices of authoritative scientific and legal institutions. He demonstrates that, as in my own analysis, the line of demarcation between fact and value is respected in each national settlement, but where that line is drawn, and who draws it, differ across the three.

The next three chapters examine from different angles another of the book's central propositions: that biological technologies interact with the law to produce new subjects and new rights. Italy features again in chapter 5, Ingrid Metzler's analysis of the politics of human embryonic stem cell (hESC) research in that country. Whereas Testa and I focus on the creation and naming of new entities as sites of constitutional inventiveness, Metzler attends to the fate of stem cells as they are caught up in the dynamics of Italian constitutional politics. In part, the controversy she describes centers on the role of the Catholic Church in appropriating as "souls" within its jurisdiction the spare embryos already created in Italy, and then preventing other similarly ambiguous entities from coming into being. In part, it is about the different incorporation of rights into bodies—the speaking bodies of genetically ill activists and the silent "bodies" of hESCs. Curiously, the bodies who spoke autonomously for themselves failed to garner enough support to overturn Italy's highly restrictive assisted reproduction law in a national referendum. Metzler shows how political abstinence (people not going to the polls) reinforced the position of a church whose injunctions of sexual abstinence were seemingly too well heeded in a nation with a famously declining birth rate.

Leaving aside borderline entities such as embryos and stem cells, Jay Aronson and David Winickoff turn to a category of troublesome human subjects—convicted criminals—and ask whether and how new biotechnologies have affected their rights under the U.S. Constitution. In chapter 6, Aronson shows how the arrival of DNA profiling, with its special claims to infallibility, intersected with habeas corpus claims, especially in death penalty cases. He inquires whether a technological advance can trigger the recognition of a new constitutional right—in this case, a right to postconviction DNA testing. Analyzing case law up to the 2009 Supreme Court decision that drew down the curtain on such a right,[5] at least for the moment, Aronson traces how a legal and a scientific debate developed together, each affecting the other. First, how foolproof is DNA typing? Second, does constitutional liberty demand that an infallible technology of truth telling, a "revelation machine," must be made available to criminal defendants? Cutting against liberal inclinations, the conservative majority's 5–4 ruling that there is no constitutional right to a DNA test displayed, in a way, a coproductionist sensibility. The justices saw the reliability of the technology, the defendants' legal strategies, and the norms of constitutional entitlement as too fluid to resolve with the bright line of a definitive constitutional settlement.

In chapter 7, Winickoff discusses how another new technology, the Combined DNA Index System (CODIS) database, forced U.S. courts to rethink their interpretation of the Fourth Amendment's protection against unreasonable search and seizure. He argues that judicial imaginaries of technology—both the forensic DNA database and technology in general—played a central role in determining doctrinal choices across several courts. Confronted by new technology, courts construed due process against their own prior understandings of what is at risk and who is entitled to be protected. In this case, that process de- and reconstructed the previously naturalized category of felons. The technology of databasing forced courts to rethink the nature and rights of this group in relation to those of the general public. Do all human subjects belong to the same class for purposes of Fourth Amendment protection; that is, do privacy rights attach in the same way to all humans, irrespective of whether they have run afoul of the law? Or should courts recognize that, with respect to rights, people may need to be differentiated on the basis of the risks they pose to society? In showing how courts differed in their responses to this question, Winickoff also establishes the indeterminacy of rights at a time when bio and information technologies are producing unprecedented intrusions into human lives.

Rights, as several of the chapters argue, emerge and are held in place in different ways in different national settings. This is partly a consequence of differences in legal traditions and governance practices from country to country, but partly also of the informal ways in which rights are built into political life worlds, that is, into the collective experiences that tie citizens to their states and vice versa. This embeddedness of rights in political culture is further explored in the next group of chapters.

In chapter 8, Mariachiara Tallacchini discusses four national models for coming to terms with the risks and promises of xenotransplantation—in the United States, the European Union, Canada, and Australia. Each political system has grappled with questions such as who can be a transplant patient; who needs to consent; who needs to be protected; what is at risk;

and what rights patients and populations have in relation to one other and to the researchers responsible for their treatment. The questions may be the same across all political systems, but the answers vary. The differences Tallacchini observes are rooted in different underlying models of state-society relations: a public health model in the European Union that weighs risks to the individual against risk to populations; a citizen model in the United States that balances individual against collective rights; and a communitarian model in Canada and Australia that allows the state to take a back seat while citizens develop collective norms through state-sponsored participatory exercises.

On a geographically orthogonal axis, contrasting the global North with the global South, Kaushik Sunder Rajan asks in chapter 9 how the emerging transnational market in personalized medicine is intersecting with state sovereignty and individual rights. Through an ethnographic comparison of clinical trials in India and the United States, he suggests that the "sovereign" who defines the rights at stake is primarily the nation-state in India but primarily the market in the United States. Experimental subjects' rights accordingly are defined and interpreted within a constitutional framing of state-society relations in India; by contrast, rights in the American case grow from a contractual framing of the position of the research subject as a potential consumer. Sunder Rajan explores the implications for bioethics in a regime (India) that bestows rights on citizens who are seen as subjects of a sometimes paternalistic state as opposed to one (United States) that confers rights on citizens who are seen as autonomous consumers of biomedical advances.

Current scholarship in science and technology studies rejects the idea that citizens are merely passive objects of the state's top-down regulation of life, or biopower. The final three chapters develop a compelling argument that the life science and technologies are sites for the articulation of new forms of political agency that can make new political rights appear where none previously existed.

In chapter 10, Jenny Reardon introduces the theme of bottom-up agency with an account of struggles for authority between socially and scientifically constituted groups of genetic research subjects. She shows that people experience group affiliation from several different positions: for example, membership in acknowledged political or social communities; capacity to assert sovereignty; and subjective identification with a community (especially for indigenous people). Whatever the basis for their sense of community, members of social groups recognize that externally imposed biological definitions of groupness may detract from the rights they enjoy through forms of group identification that have no genetic markers attached to them. Reardon demonstrates how contests over participation in research may pit these two notions of belonging to groups—the biological and the social—against one another.

In chapter 11, Robert Doubleday and Brian Wynne examine a series of encounters between UK citizens and the state at the turn of the twentieth century, centering largely on the introduction of genetically modified crops and foods into Britain. Citizens, they argue, initiated a form of "uninvited participation" through various types of direct and symbolic action, thereby redefining in effect the participatory rights that citizens should enjoy in the development of national imaginaries of agricultural biotechnology, innovation, and progress. Their chapter looks at counterpoised tendencies in this contested period of British politics toward opening up and closing down the possibilities for citizen engagement—the former representing the continuance of ancient monarchical practices of governance, the latter a potentially new reconstitutionalizing of the British subject in relation to the technoscientific state.

In chapter 12, Jim Dratwa takes a still wider-angled look at the construction of participatory rights with a close reading of the European Union as a political space committed to safeguarding the lives of its citizens through precautionary policy making. In his account, the European Parliament and Commission move from a bureaucratic-rationalistic mode of self-legitimation (based on technical risk assessment) to one that foregrounds technological uncertainty and positions European institutions, as opposed to those of member states, as the ones best equipped to govern the uncertainties of new technologies. Dratwa shows how the European Parliament's textual invocation of "precaution" as a governing principle—in relation to biotechnology among other hazards—calls into being a particular vision of Europe, along with a European public that gives assent to governance (and thus acquires rights) at the European level. He demonstrates how European bodies, through linguistic and procedural choices, mediate among different national positions, thereby giving rise to an emerging, precautionary, European subjectivity.

Conclusion

Several decades of development in the life sciences and technologies have initiated wide-ranging interactions between scientific and legal, particularly constitutional, orders. Sometimes explicitly and formally constitutional, as in cases involving U.S. criminal defendants' rights to genetic information, and sometimes hidden beneath contestation over sovereignty or representation, as in U.S. diversity research, Indian pharmaceutical trials, and UK and EU participatory processes, new ways of knowing life through genetic texts have opened the way to rewriting principles of individual and collective rights. At the same time, a multiplicity of new, technologically created biological entities are raising questions about the relevance of already well-recognized rights to things derived from human bodies, things with the potential to become human, and things combining human and nonhuman characteristics. These developments add up to a hitherto largely undiscussed process of bioconstitutionalism that may fundamentally redefine the natural law–centered concepts of rights inherited from the democratic revolutions of the eighteenth century.

As is clear in the following chapters, two central questions for law and ethics continually resurface in the era of technologically manipulable life: who belongs to communities of moral concern, and who is responsible for taking care of life in those communities? For stem cells and chimeras, patients and prisoners, research subjects and consumers of genetically modified crops, ontologies, classifications, and rights have all been redefined through novel intersections between the texts of law and of life. The resulting multiplicity of readings—of entities, entitlements, and responsibilities—has put on the table new questions of stewardship and sovereignty. Nation states emerge in our accounts as prime sites for working out the constitutional challenges raised by these events. Denying any purely mentalist or static conceptions of national culture, we show through detailed empirical analysis how institutionalized values and practices shape the territories in which life itself gains meaning and constitutional norms are reworked. Morally, ethically, legally, and scientifically these are interesting times. Readers will find in the following pages more than mere tea leaves for reading our bioconstitutional future.

Notes

1. Here and in the volume as a whole, "coproduction" refers to the concurrent formation of natural and social orders in societies with substantial investments in scientific and technological innovation. The concept is widely used in science and technology studies to describe the complex linkages forged in modernity between facts and values, descriptive and normative, epistemic and material (for further discussion, see Jasanoff 2004).

2. Britain's Human Fertilisation and Embryology Authority (HFEA), for example, approved the procedure, provided that prenatal screening also increased the selected embryo's chances of being born disease-free. The House of Lords ratified the

HFEA's decision (*Quintavalle v. Human Fertilisation and Embryology Authority* [2005] UKHL 28).

3. Democratic constitutions were first imagined more than two hundred years ago by representatives of preindustrial, agrarian societies, situated mainly in Europe and North America. Since that time, innumerable changes have occurred in the organization of commerce and industry, among them radical shifts in transportation, communication, financial, agricultural, medical, and manufacturing practices. The society known to the lawyers, merchants, plantation owners, and politicians who drafted the Declaration of Independence and the U.S. Constitution no longer exists. Ackerman (1983) argues that constitutional law, too, has changed profoundly to meet the challenges of industrial development, but these changes have not always taken place through constitutional amendment—witness, for example, the wholesale transformation of administrative legal culture in the New Deal.

4. *Association for Molecular Pathology, et al. v. United States Patent and Trademark Office*, 702 F. Supp. 2d 181 (S.D.N.Y. 2010).

5. *District Attorney's Office for Third Judicial District v. Osborne*, 557 US (2009).

References

Ackerman, Bruce A. 1983. *Reconstructing American Law*. Cambridge, Mass.: Harvard University Press.

Árnason, Vilhjálmur, and Stefán Hjörleifsson. 2007. Geneticization and Bioethics: Advancing Debate and Research. *Medicine, Health Care, and Philosophy* 10 (4): 417–431.

Aronson, Ori. 2009–2010. Inferiorizing Judicial Review: Popular Constitutionalism in Trial Courts. *University of Michigan Journal of Law Reform* 43:971–1049.

Bartlett, Katherine T. 1990. Feminist Legal Methods. *Harvard Law Review* 103:829–888.

Berg, Paul, David Baltimore, Sydney Brenner, Richard O. Roblin, III, and Maxine F. Singer. 1975. Summary Statement of the Asilomar Conference on Recombinant DNA Molecules. *Proceedings of the National Academy of Sciences of the United States of America* 72 (6): 1981–1984.

Bijker, Wiebe, Thomas Hughes, and Trevor Pinch. 1987. *The Social Construction of Technological Systems*. Cambridge, Mass.: MIT Press.

Boyle, James. 1992. A Theory of Law and Information: Copyright, Spleens, Blackmail, and Insider Trading. *Stanford Law Review* 80:1508–1520.

Breyer, Stephen. 1993. *Breaking the Vicious Circle: Toward Effective Risk Regulation*. Cambridge, Mass.: Harvard University Press.

Buchanan, Allen E., Dan W. Brock, Norman Daniels, and Daniel Wikler. 2000. *From Chance to Choice: Genetics and Justice*. Cambridge: Cambridge University Press.

Callon, Michel. 1986. Some Elements of a Sociology of Translation: Domestication of the Scallops and the Fishermen of St. Brieuc Bay. In *Power, Action, and Belief: A New Sociology of Knowledge?*, ed. John Law, 196–233. New York: Routledge.

Cambrosio, Alberto, et al. 1990. Scientific Practice in the Courtroom: The Construction of Sociotechnical Identities in a Biotechnology Patent Dispute. *Social Problems* 37:275–293.

Duster, Troy. 1990. *Backdoor to Eugenics*. New York: Routledge.

Epstein, Steven. 2007. *Inclusion: The Politics of Difference in Medical Research*. Chicago: University of Chicago Press.

Ezrahi, Yaron. 1990. *The Descent of Icarus: Science and the Transformation of Contemporary Democracy*. Cambridge, Mass.: Harvard University Press.

Fisher, William W., III, Morton J. Horwitz, and Thomas A. Reed, eds. 1993. *American Legal Realism*. New York: Oxford University Press.

Foster, Kenneth, and Peter Huber. 1997. *Judging Science: Scientific Knowledge and the Federal Courts*. Cambridge, Mass.: MIT Press.

Foucault, Michel. 1998. *The Will to Knowledge*. Vol. 1. The History of Sexuality. London: Penguin.

Fried, Barbara. 1998. *The Progressive Assault on Laissez Faire: Robert Hale and the First Law and Economics Movement*. Cambridge, Mass.: Harvard University Press.

Fukuyama, Francis. 2002. *Our Posthuman Future: Consequences of the Biotechnology Revolution*. New York: Farrar, Straus and Giroux.

Goldberg, Steven. 1994. *Culture Clash*. New York: New York University Press.

Gottweis, Herbert. 1998. *Governing Molecules: The Discursive Politics of Genetic Engineering in Europe and the United States*. Cambridge, Mass.: MIT Press.

Haraway, Donna. 1991. *Simians, Cyborgs, and Women: The Reinvention of Nature*. New York: Routledge, Chapman, and Hall.

Haraway, Donna. 1997. *Modest-Witness@Second-Millennium.FemaleMan-Meets-Oncomouse: Feminism and Technoscience*. New York: Routledge.

Harding, Sarah. 2006. Kramer's *Popular Constitutionalism*: A Quick Normative Assessment. *Chicago Kent Law Review* 81:1117–1126.

Hart, H. L. A. 1961. *The Concept of Law*. Oxford: Clarendon Press.

Häyry, Matti, Ruth Chadwick, Vilhjálmur Árnason, and Gardar Árnason. 2007. *The Ethics and Governance of Human Genetic Databases: European Perspectives*. New York: Cambridge University Press.

Hilgartner, Stephen, and Sherry I. Brandt-Rauf. 1994. Data Access, Ownership, and Control. *Knowledge* 15:355–372.

Hindmarsh, Richard, and Barbara Prainsack, eds. 2010. *Genetic Suspects: Global Governance of Forensic DNA Profiling and Databasing*. Cambridge: Cambridge University Press.

Hotz, Robert L. 2010. "Scientists Create Synthetic Organism." *Wall Street Journal*, May 21.

Huber, Peter W. 1991. *Galileo's Revenge: Junk Science in the Courtroom*. New York: Basic Books.

Jasanoff, Sheila. 2008. Making Order: Law and Science in Action. In *Handbook of Science and Technology Studies*, 3rd ed., ed. Edward Hackett, et al., 761–786. Cambridge, Mass.: MIT Press.

Jasanoff, Sheila. 2005. *Designs on Nature: Science and Democracy in Europe and the United States*. Princeton: Princeton University Press.

Jasanoff, Sheila. 2003. In a Constitutional Moment: Science and Social Order at the Millennium. In *Social Studies of Science and Technology: Looking Back, Ahead*, Sociology of the Sciences Yearbook 23, ed. Bernward Joerges and Helga Nowotny, 155–180. Dordrecht: Kluwer.

Jasanoff, Sheila. 2001. Ordering Life: Law and the Normalization of Biotechnology. *Politeia* 17 (62): 34–50.

Jasanoff, Sheila. 1995. *Science at the Bar: Law, Science, and Technology in America*. Cambridge, Mass.: Harvard University Press.

Jasanoff, Sheila. 1990. *The Fifth Branch: Science Advisers as Policymakers*. Cambridge, Mass.: Harvard University Press.

Jasanoff, Sheila. 1987. Biology and the Bill of Rights: Can Science Reframe the Constitution? *American Journal of Law & Medicine* 13:249–289.

Jasanoff, Sheila, and Sang-Hyun Kim. 2009. Containing the Atom: Sociotechnical Imaginaries and Nuclear Regulation in the U.S. and South Korea. *Minerva* 47 (2): 119–146.

Jasanoff, Sheila, ed. 2004. *States of Knowledge: The Co-Production of Science and Social Order*. London: Routledge.

Jasanoff, Sheila, Gerald Markle, James Petersen, and Trevor Pinch, eds. 1995. *Handbook of Science and Technology Studies*. Newbury Park, Calif.: Sage Publications.

Jordan, Kathleen, and Michael Lynch. 1999. The Dissemination, Standardization, and Routinization of a Molecular Biological Technique. *Social Studies of Science* 28 (5–6): 773–800.

Kairys, David, ed. 1990. *The Politics of Law*. New York: Pantheon.

Kay, Lily E. 2000. *Who Wrote the Book of Life? A History of the Genetic Code*. Palo Alto: Stanford University Press.

Kay, Lily E. 1993. *The Molecular Vision of Life*. New York: Oxford University Press.

Keasling, Jay. 2005. The Promise of Synthetic Biology. *Bridge* 35 (4): 18–21.

Keller, Evelyn Fox. 1992. Nature, Nurture, and the Human Genome Project. In *The Code of Codes*, ed. Daniel J. Kevles and Leroy Hood, 281–299. Cambridge, Mass.: Harvard University Press.

Keller, Evelyn Fox. 2000. *The Century of the Gene*. Cambridge, Mass.: Harvard University Press.

Kelman, Mark. 1987. *A Guide to Critical Legal Studies*. Cambridge, Mass.: Harvard University Press.

Kevles, Daniel J. 1985. *In the Name of Eugenics: Genetics and the Uses of Human Heredity*. Cambridge, Mass.: Harvard University Press.

Kevles, Daniel J., and Leroy Hood, eds. 1992. *The Code of Codes*. Cambridge, Mass.: Harvard University Press.

Kitcher, Philip. 1996. *The Lives to Come*. New York: Simon and Schuster.

Kramer, Larry D. 2004. *The People Themselves: Popular Constitutionalism and Judicial Review*. New York: Oxford University Press.

Kysar, Douglas. 2010. *Regulating from Nowhere: Environmental Law and the Search for Objectivity*. New Haven: Yale University Press.

Latour, Bruno. 1987. *Science in Action*. Cambridge, Mass.: Harvard University Press.

Latour, Bruno. 1988. *The Pasteurization of France*. Cambridge, Mass.: Harvard University Press.

Latour, Bruno. 1993. *We Have Never Been Modern*. Cambridge, Mass.: Harvard University Press.

Latour, Bruno, and Steve Woolgar. 1979. *Laboratory Life: The Construction of Scientific Facts*. Princeton: Princeton University Press.

Lessig, Lawrence. 1997. Constitution of Code: Limitations on Choice-Based Critiques of Cyberspace Regulation. *CommLaw Conspectus* 5:181–204.

Lewontin, Richard. 1993. *Biology as Ideology*. New York: Harper Perennial.

MacKenzie, Donald. 1990. *Inventing Accuracy: A Historical Sociology of Nuclear Missile Guidance*. Cambridge, Mass.: MIT Press.

Mol, Annemarie, and John Law. 1994. Regions, Networks and Fluids: Anaemia and Social Topology. *Social Studies of Science* 24:641–671.

Petryna, Adriana. 2002. *Life Exposed: Biological Citizens after Chernobyl*. Princeton: Princeton University Press.

Porter, Theodore M. 1995. *Trust in Numbers: The Pursuit of Objectivity in Science and Public Life*. Princeton: Princeton University Press.

Post, Robert, and Reva Siegel. 2007. *Roe* Rage: Democratic Constitutionalism and Backlash. *Harvard Civil Rights-Civil Liberties Law Review* 42:373–433.

Rabinow, Paul. 1992. Artificiality and Enlightenment: From Sociobiology to Biosociality. In *Incorporations*, ed. Jonathan Crary and Sanford Kwinter, 234–252. New York: Bradbury Tamblyn and Boorne Ltd.

Rose, Nikolas. 2006. *The Politics of Life Itself: Biomedicine, Power, and Subjectivity in the Twenty-First Century*. Princeton: Princeton University Press.

Sandel, Michael. 2007. *The Case against Perfection: Ethics in the Age of Genetic Engineering*. Cambridge, Mass.: Harvard University Press.

Schauer, Frederick. 1998. Internet Privacy and the Public-Private Distinction. *Jurimetrics* 38:555–564.

Schuck, Peter. 1993. Multi-Culturalism Redux: Science, Law, and Politics. *Yale Law & Policy Review* 11:1–46.

Shapin, Steven. 1994. *A Social History of Truth*. Chicago: University of Chicago Press.

Shapin, Steven, and Simon Schaffer. 1985. *Leviathan and the Air-Pump: Hobbes, Boyle, and the Experimental Life.* Princeton: Princeton University Press.

Silver, Lee M. 1997. *Remaking Eden: How Genetic Engineering and Cloning Will Transform the American Family.* New York: Harper Collins.

Smith, Merritt Roe, and Leo Marx, eds. 1994. *Does Technology Drive History? The Dilemma of Technological Determinism.* Cambridge, Mass.: MIT Press.

Smith, Roger, and Brian Wynne, eds. 1989. *Expert Evidence: Interpreting Science in the Law.* London: Routledge.

Specter, Mark. 2009. A Life of Its Own. *New Yorker* (September): 28.

Star, S. Leigh, and James Griesemer. 1989. Institutional Ecology, "Translations" and Boundary Objects: Amateurs and Professionals in Berkeley's Museum of Vertebrate Zoology, 1907–1939. *Social Studies of Science* 19:387–420.

Stone, Christopher D. 1974. *Should Trees Have Standing? Toward Legal Rights for Natural Objects.* Los Altos, Calif.: William Kaufmann.

Sunstein, Cass. 1993. *The Partial Constitution.* Cambridge, Mass.: Harvard University Press.

Waldron, Jeremy. 1999. *Law and Disagreement.* New York: Oxford University Press.

Watson, James D., and Francis H. C. Crick. 1953. Molecular Structure of Nucleic Acids: A Structure for Deoxyribonucleic Acid. *Nature* 171:737–738.

Wright, Susan. 1994. *Molecular Politics: Developing American and British Regulatory Policy for Genetic Engineering, 1972–1982.* Chicago: University of Chicago Press.

Wynne, Brian. 1982. *Rationality and Ritual: The Windscale Inquiry and Nuclear Decisions in Britain.* Chalfont St. Giles, UK: British Society for the History of Science.

Wynne, Brian. 1988. Unruly Technology. *Social Studies of Science* 18:147–167.

2

States of Eugenics: Institutions and Practices of Compulsory Sterilization in California

Alex Wellerstein

Between 1909 and the early 1950s, the state of California sterilized over twenty thousand patients in government institutions for the mentally ill and mentally deficient. Of the many states that had compulsory sterilization programs, California's was by far the largest in terms of patients sterilized, affecting nearly as many people as the sum of the totals from the next four top-sterilizing states combined (figure 2.1).[1] The motivation for these sterilizations has traditionally been associated with the concept of eugenics: the desire to improve the human gene pool by discouraging the reproduction of the "unfit." These mass sterilizations have generally been taken as the most tangible and permanent of all of the American forays into eugenics, and its closest link to the genocidal policies practiced by National Socialist Germany.

The history of eugenics is generally told explicitly as "a history of a bad idea" (e.g., Carlson 2001). It is an intellectual history, an account of the dangerous power of ideology-infected science. This framework, which dominated historical accounts of eugenics since they first started being written in the 1960s, tended to focus on the genesis and transformation of eugenic thought as reflected in the writings of eugenic propagandists and occasionally state legislation.[2] Aside from legislation for immigration restriction, eugenics had very little federal recognition in the United States, and was prosecuted mainly on a state-by-state basis. In the American case, the intellectual history approach has been used extensively to trace strong connections between the American embrace of eugenics and the case of the Nazis (Black 2003). Such comparisons pack considerable rhetorical impact in a culture that has long prided itself on its crusading role in the Second World War, and the history of eugenics and possible eugenic futures have become the standard case study of the intersection between biology and society.

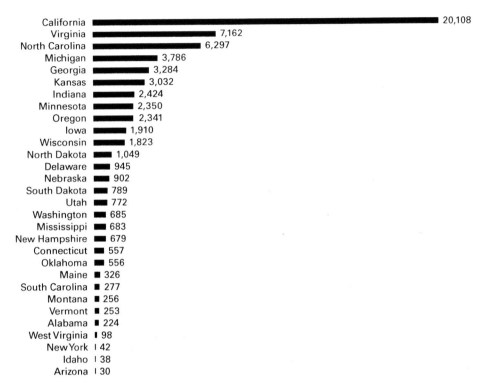

Figure 2.1
Cumulative sterilizations by state, 1907–1964, out of 63,643 total. These statistics are mis-
leadingly precise (many sterilizations no doubt went unreported), but the order of magnitude
is probably correct. *Source*: Robitscher 1973, appendix 2; graph by author.

But does this top-down, idea-centric view actually illuminate the American case? I argue in this chapter that California's history of sterilization
shows that it does not. A history of ideology neither explains why California's sterilization rates were so much higher than the rest of the country
nor gives an account of why they dropped off dramatically in the early
1950s. Although it was the most influential of the state sterilization stories—the Nazis famously pointed to California's success when embarking
on their own mass sterilization program (Kühl 1994, 39–44)—California's
sterilization program has generally been lumped in with the overall story
of American eugenics in a way that neither recognizes nor explains its
particularities.[3] In this chapter, I look closely at the institutional, organizational basis of sterilization in California, tracing how the power to
sterilize—and the questions of who to sterilize, why, and perhaps why
not—wended their way through legal, medical, and local frameworks.

Because of the particulars of the California case, power ended up being disproportionately concentrated in the hands of individual hospital administrators, who were often intellectually and physically quite distant from the direct influence of eugenics. This institutional view of eugenics paints a more subtle picture of the ways in which ideology undergoes translation and transformation as it becomes practice. The particular model for that process here is one that may call for a more general reevaluation of our overall understanding of eugenics in the American context.[4]

This historical chapter contributes on several levels to a broader work on our contemporary bioconstitutional moment. First, the history of eugenics has been *the* primary lens through which questions of biological power have been read in the late twentieth and early twenty-first century. When James D. Watson decided that the Human Genome Project should, from the very beginning, devote considerable funds to ethical, legal, and social questions, it was because he decided that his failure to do so "might falsely be used as evidence that I was a closet eugenicist" (Watson 2001, 206). Further, our understandings of what is important and what is at stake in questions about the intersections of biology, law, and society are heavily rooted in our historical understanding of past wrongs and their origins. In the case of sterilization laws, the exceptional focus on "ideas" as the motive force has, I argue, diverted our attention from the important ways in which ideas, institutions, and practices are interwoven through processes of coproduction. This chapter adds important nuances and qualifications to the ways in which we think about the application of extreme biological power in a specifically American institutional context; by historicizing the American case, it enriches our understanding of this country's political culture and thereby contributes to the volume's comparative project. At the same time, the chapter points toward the reframing of our historical assumptions in the light of contemporary scientific and technological advances.

Sterilization in the Golden State

In 1922, the Cold Spring Harbor eugenicist Harry H. Laughlin compiled a five-hundred-page monograph devoted to the systematic study of "Eugenical Sterilization in the United States." Laughlin did this as an enthusiastic promoter of sterilization of the "unfit," yet it remains one of the most careful studies of the *implementation* of eugenics in the United States. Laughlin was an unreliable narrator with respect to the scientific benefits of eugenic sterilization, but as a reporter of the state of legislative and political situation in the early 1920s, he displayed a sharp eye for policy analysis (Laughlin 1922; Kevles [1985] 1995, 108–118).

For an avid advocate of eugenic sterilization in the United States, it was not a good moment. There was no federal statute regulating sterilization, and the prospects of one were slim to none. Sterilization laws were adopted exclusively at the state level and were primarily intended for implementation within state-run institutions. By early 1921, fifteen states had passed sterilization statutes, but five of those laws had been struck down as unconstitutional by state courts, and one state had repealed its law. Five other states had pushed sterilization laws through their legislatures, only to have them vetoed by the governor or revoked by popular referendum. Of the nine states that still had sterilization laws on the books, only two—California and Nebraska—seemed to function well administratively, and only California was sterilizing to any great effect (almost 80 percent of the reported 3,233 nationwide sterilizations were performed in the state) (Laughlin 1922, 96–97).

In Laughlin's eyes there were multiple problems. Chief among these were the sterilization statutes themselves. Sterilization was considered a controversial enough operation that it could not simply be justified as a normal operating procedure—unlike an appendectomy, for example, the operation was not deemed as being a medical necessity for the patient, and it required special authorizing legislation. Poorly written statutes created all sorts of difficulties. Physicians at state hospitals were afraid to rely on laws that might be ruled unconstitutional by the courts, as they or their hospitals could then be held liable for mayhem or malpractice. Even in states where statutes had not been challenged, the fear that they might be challenged was significant, as physicians pointed out in testimonials to Laughlin. In states where constitutionality was not a major concern, physicians reported that the administration of the law was so convoluted or contradictory that they were effectively prevented from carrying out operations by excessive "red tape" (Laughlin 1922, 52–92). This was, in the end, Laughlin's impetus for crafting his own "Model Eugenical Sterilization Law" as a guide to state legislators, and to push immediately for a test case to prove its validity. The result was the infamous U.S. Supreme Court judgment in *Buck v. Bell* (1927), in which Justice Oliver Wendell Holmes, Jr., enthusiastically accepted the argument that compulsory sterilization was no more severe a public health measure than compulsory vaccination (Laughlin 1922, 445–452; Lombardo 1985, 2008).

Laughlin had mixed feelings about California. On the one hand, he found its enthusiastic embrace of sterilization encouraging: "To California must be given the credit for making the most use of her sterilization laws. The history of the application of these statutes shows an honest and competent effort to improve 'the racial qualities of future generations'" (Laughlin 1922, 52). California seemed to suffer from none of the legal complications that plagued other states. Its sterilization statutes had been revised numerous times since first enacted in 1909, and California's long-serving Attorney General, Ulysses S. Webb, had explicitly endorsed the constitutionality of sterilization soon after it had appeared (Webb 1910). But if "red tape" was not an issue, there was another administrative problem.

Laughlin was well aware that it was not just an "honest and competent effort" that accounted for the actual implementation of state laws. In his analysis, a secure legal environment was required (hence the need for a "Model Law"), and it needed to be implemented in an institutional environment that would not overly complicate it with red tape, contradictory requirements, or other poor "administrative machinery." Laughlin was no jurist, but he took pains to distinguish between "mandatory and optional elements" of sterilization laws—where "mandatory" and "optional" referred to requirements for the physician, not the patient. If laws gave physicians too much discretion as to whom and why to sterilize, the results would be haphazard:

If a law is meant to be compulsory [for the physicians], then of course there must be no gaps in its chain of mandates, which begins with the order for the appointment of executive officers, and ends with the actual surgical operation of sterilization. A single "may" inserted in the chain of execution makes the whole procedure an optional, or at least a non-compulsory one. The principal elements in the chain are: (1) the appointment of executive agents; (2) the examination of individuals alleged to be subject to the act; (3) the determination of the facts in particular cases, whether the particular person is subject to eugenical sterilization; (4) the order for the actual sterilizing operation. (Laughlin 1922, 114–117)

California got very low marks when judged by this standard. The state sterilization law contained seven "mays," which Laughlin highlighted in bold type: its "chain of mandates" had considerable gaps. Despite its high rate of use, Laughlin considered the statute "ineffective" because it gave the superintendents of individual hospitals far too much freedom to implement the law or to disregard it. As a psychologist who later opposed the sterilization law put it: "They were not ordered to sterilize—they were permitted to sterilize" (Tarjan 1998, 227).

Laughlin's irritation provides a useful analytic lens. California enacted two laws in 1909 and 1913 under which almost all sterilizations were performed until the rates suddenly plummeted after 1949 and went to near zero after 1951.[5] The 1909 statute consisted of a single paragraph:

Whenever in the opinion of the medical superintendent of any state hospital, or the superintendent of the California Home for the Care and Training of Feeble-Minded Children, or of the resident physician in any state prison, it would be beneficial and conducive to the benefit of the physical, mental or moral condition of any inmate of said state hospital, home, or state prison, to be asexualized, then such superintendent or resident physician shall call in consultation the general superintendent of state hospitals and the secretary of the state board of health, and they shall jointly examine into all the particulars of the case with the said superintendent or resident physician, and if in their opinion, or in the opinion of any two of them, asexualization will be beneficial to such inmate, patient or convict, they may perform the same.[6]

"Whenever in the opinion" is the crucial phrase that defines the character of sterilization in California: operations were ordered *at the discretion* of hospital superintendents. Though the laws would change, this fundamental delegation of judgment would not.

The 1909 statute, as noted, did not specify the motivation for sterilization operation too finely; it needed to be only "beneficial and conducive to the benefit of the physical, mental or moral condition" of the patient—a vague requirement centered around value to the *individual patient*, not to any notion of a collective "germ plasm," "gene pool," or "future stock," as eugenicists might have wished. Even the term "asexualization" is vague, being easily associated with castration (an operation with which eugenicists generally did not want their cause to be associated).[7]

The law's perceived vagueness led to its speedy repeal and replacement with a new, longer sterilization law only four years later. The 1913 statute provided that the centralized bureaucracy that administered the mental hospitals (the State Commission in Lunacy, whose name changed successively to the Department of Institutions and then the Department of Mental Hygiene) could, at its discretion, sterilize a patient. This section seemed to change the lines of authority considerably, but in practice there is no evidence that the centralized bureaucracy went out of its way to identify and order sterilizations without being requested by a hospital superintendent. (Many decades later, California's Director of Mental Hygiene noted that "this provision of law has not been followed since its enactment."[8]) Importantly, the first section of the 1913 statute specified that the patient must be "afflicted with hereditary insanity or incurable chronic mania or dementia," introducing heredity into the determination for the first time, and thus providing concrete evidence that eugenics was a consideration.[9] Yet the law's second section was simply an exact duplicate of the original

1909 statute, which enabled sterilization at the discretion of physicians, with no further specification of the kinds of reasons that had to be given.

Additional statutes enacted in 1917 and 1923 changed some of the grounds for sterilization. The 1917 statute added "those suffering from perversion or marked departures from normal mentality or from disease of a syphilitic nature" to the classes of persons who could be sterilized, and the 1923 statute specified that prisoners who had committed sexual abuse on girls under the age of ten could be sterilized "for the prevention of procreation." Neither revision, however, changed the lines of command or refined the reasons why a medical superintendent could request sterilization.[10] No further changes to this legislation took place until 1951, when the law was substantially rewritten as part of a general overhaul of mental health legislation.[11]

After the 1913 revision, then, California law allowed for sterilization of hospital inmates for a variety of reasons, including both what could be considered eugenic grounds (heredity) and what could plausibly be considered therapeutic grounds (benefits to individual patients), as well as provisions for punitive sterilizations of prisoners who had committed certain sexual crimes. And the law allowed both the state hospital administration and individual hospital superintendents to determine who would be a candidate for sterilization, although in practice only the latter recommended patients for sterilization.

California, like many other states, had an enthusiastic eugenicist lobby in the form of the Human Betterment Foundation, founded by financier E. S. Gosney with the aid of biologist Paul B. Popenoe, in 1928. Gosney and Popenoe had been involved in tracking the progress of sterilization in California since 1925, and had been using it to proselytize elsewhere (including, notably, Germany) (Popenoe 1934; Stern 2005, chap. 5). But these eugenicists, however eager, were disconnected from the actual practice of sterilization: they corresponded with state superintendents primarily in seeking information for their reports (Gosney and Popenoe, 1929). Although they actively set up networks of allies, as well as attempts to influence both public and private opinion in favor of sterilization, they were ultimately outsiders (Stern 2005, chap. 5). This is not to say that Gosney, Popenoe, and the other eugenic "propagandists" were completely powerless—at times they enjoyed wide popular support, as most histories of this period document. But they had no power over the operations of individual mental institutions or of the medical system as a whole. Wide influence on thought and discourse is a very different thing from the translation of ideas into concrete, localized practice.

If one followed Laughlin's model law, eugenic ideology would have been written into the text of the law itself. In the California case, this clearly did not occur. The original 1909 text of the law was drafted and encouraged by the Secretary of the State Lunacy Commission, Dr. Frederick W. Hatch, Jr., a physician with definite eugenic inclinations. Until his death in 1924, he retained the ability to approve or veto suggestions for sterilization from the state superintendents, but he could not nominate candidates himself. The examples of sterilization requests approved by Hatch that Laughlin includes in his study do not show Hatch exercising the statutory discretion he possessed: he approves all requests, whether they indicate heredity as a factor or not (Laughlin 1922, 52–53). The law he wrote may have had eugenics as one of its motivations, but it did not impose those motivations on physicians. To understand the practice of sterilization, then, we must look away from the eugenicists, and even the expressed intentions of the law, and turn to the practices of the superintendents themselves.

Superintendents' Views on Sterilization

When the first sterilization law came into effect in 1909, California had five state hospitals for the mentally ill (Agnews, Mendocino, Napa, Patton, and Stockton State Hospitals) and one home for the mentally deficient (Sonoma State Home). All but Patton were in the northern part of the state. In 1916 and 1917, respectively, additional hospitals for the mentally ill (Norwalk State Hospital) and the mentally deficient (Pacific Colony) were founded in southern California, to help accommodate the increasing demand for state mental hygiene resources. Three more hospitals for the mentally insane were added in the late 1930s through the 1940s (Camarillo, DeWitt, and Modesto State Hospitals), adding up by the end of California's sterilization period to a total of nine hospitals for the mentally insane and two homes for the mentally deficient in which sterilizations were performed (figure 2.2).[12]

Operations were not equally distributed among these hospitals. By 1950, the last year in which hospital-by-hospital sterilization rates are available, three institutions alone accounted for 68 percent of all sterilizations performed (table 2.1). Though some differences can be attributed to the small size of a number of the institutions (DeWitt and Modesto had both just become operational by the late 1940s), on the whole the population differences were not large enough to account for the disparities in sterilization numbers. Camarillo, for example, had eclipsed all other institutions in patient population by the 1940s, yet it accounts for an

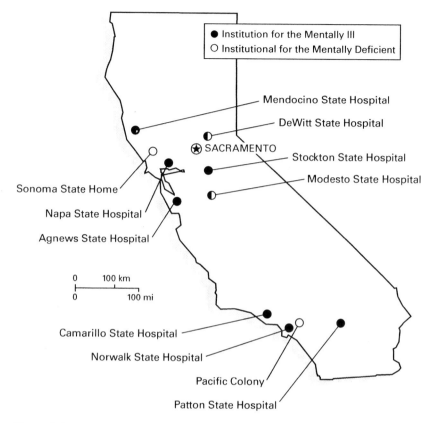

Figure 2.2
California's state mental health system for the primary period of compulsory sterilization as of 1950. *Source*: Adapted by author from California Department of Mental Hygiene 1950, chart I, 3.

insignificant fraction of the total sterilizations. Agnews always had considerably more patients than Norwalk, Pacific Colony, and Sonoma, yet it sterilized considerably fewer than these.

Three institutions—Stockton, Sonoma, and Agnews—vividly demonstrate the different varieties of sterilization practice that flourished within an institutional model that stressed the autonomy and the discretion of hospital superintendents. I have chosen these not because they are necessarily representative of all sterilizing institutions (though together they account for nearly 50 percent of all sterilizations performed in the state), but because they illustrate the almost limitless power that their long-tenured administrators wielded with regard to sterilization policy.

Table 2.1
Cumulative sterilizations at California state institutions, 1909–1950

Sonoma	5,530	29.4%
Patton	4,585	24.4%
Stockton	2,669	14.2%
Napa	1,843	9.8%
Pacific Colony	1,759	9.4%
Norwalk	1,167	6.2%
Agnews	799	4.3%
Mendocino	364	1.9%
Camarillo	58	0.3%
DeWitt	15	0.1%
Modesto	3	0.0%
Total	18,792	100%

Source: Compiled from California Department of Mental Hygiene 1950, tables 60 and 121, on 142 and 239.

Stockton State Hospital

The historian of medicine Joel Braslow provides the most definitive study of sterilization practices at Stockton State Hospital (Braslow 1996; Braslow 1997, chap. 3). The oldest such institution in California, Stockton operated from 1906 until 1929 under the directorship of Dr. Fred P. Clark, who, quite unusually for the period, had inherited this mantle from his father. In both his published writings and, as Braslow has shown, in his administrative practices, Clark favored sterilization as a form of therapy. Clark wrote in a 1924 report to the governor and the other superintendents that vasectomies, in particular, had a positive effect on his patients' mental activities:

The law permitting the sterilization of the insane to my mind is one of the best things that has been done to prevent the unfit from reproducing their kind. Besides this feature of the law, in many cases of the men the operation has had a very beneficial effect upon their mental condition, that is, we have had numerous cases whose mental condition improved up to a certain point and then remained stationary. After these patients were sterilized many of them recovered completely and have had no recurrence of their mental trouble. (Clark 1924, 101–102)

Clark subscribed to what was known as the "Steinach method" of sterilization, named after the Austrian endocrinology pioneer Eugen Steinach, who had studied the supposed revitalization obtained in rats, guinea pigs, and eventually humans after vasectomies. Steinach formulated a theory that the severing of the *vas deferens* forced the testicles to increase their production of hormones, giving new energy and life to the patient. Other theories involved speculation on the effects of reabsorption of testicular fluids into the bloodstream.[13] As Clark explained in 1916:

By this interruption in the continuity of the vas, the testicular secretion is absorbed. Since performing these operations we are led to believe, by the improvement in general and mental health, there is a distinct beneficial result from the absorption of the testicular secretion. . . . Many of the results claimed [by others, in the past, for such operations] were evidently due to suggestion. However, since beginning these sterilization operations, we are led to believe that by the improvement in mental and general health that there is a definite beneficial effect from [vasectomy] and may lead to important findings as an organo therapeutic agent. (Laughlin 1922, 56)

Theories of therapeutic sterilization were not widely held in the medical community and were viewed with skepticism by most eugenicists. California sterilization advocates Gosney and Popenoe skeptically noted in 1929 that "the patient seems to get 'rejuvenation' when he expected it and paid for it; when he did not expect it, and paid merely for sterilization, he got nothing but sterilization" (Gosney and Popenoe 1929, 89). Another text on eugenic sterilization endorsed "the Steinach method" as yet another reason why sterilization was desirable for mental patients, but offered it simply as an additional benefit to counter claims that sterilization would have negative physiological effects (Landman 1932, 235–236).

It is precisely the legal and bureaucratic decentralization of sterilization in California that allowed the Steinach method to be translated into practice. This was exactly the sort of noneugenic discretion that Laughlin feared would thrive under such imprecisely worded laws. The legal arrangement invested Clark with the power to choose whom he sterilized and why—even if it was in the name of a medically fringe theory, one that other California superintendents did not embrace.[14] The anomaly of the Stockton program is exactly what makes it so revealing of the overall organization of sterilization in California, and what makes drawing a straight line from national ideologies to the idiosyncrasies of local practice so difficult.

Theories about vasectomies and testicular fluid clearly applied only to male patients. For females, Clark took a somewhat different position, one equally centered on the benefits of the sterilization to the individual, but he expressed that benefit in social rather than physiological terms. As he wrote in his 1922 biennial report, "We have sterilized quite a number of patients during the past year, both men and women. In many of the men, we have noticed very marked improvement in their mental condition after they have been sterilized. In the women it has prevented a recurrence of their mental trouble where it was due to childbirth" (Clark 1922, 88).

Was this eugenics? Braslow has categorized this motive as therapeutic, and distinguished it from eugenics because the latter "was meant to treat a sick and degenerating society, not the suffering of individuals" (Braslow 1996, 40). Wendy Kline, in contrast, has argued that such paternalism toward women patients was always present in the eugenics movement, whose notions of masculinity and femininity often led to the targeting of the "misfit mother" (Kline 2001, chap. 4). In either case, what was at work in these cases is not a simple ideology of gene purification.

Braslow's research has shown that these claims of the therapeutic value of male sterilization were not simply rhetorical. Therapeutic motives can be seen quite explicitly in private interactions between patients and physicians at Stockton, reconstructed from verbatim transcripts of patient interviews and meetings between physicians. During Clark's long tenure, explicitly hereditarian concerns played a minor role in sterilization requests by physicians at Stockton. In many cases, the sterilizations were not compulsory in the sense of being done completely against the will of patients or their family members. According to Clark, despite the compulsory option in the state sterilization laws, patient or guardian consent was usually procured: "It is not necessary to obtain the consent of the relatives of the patient or of the patient himself but it has always been our custom to obtain to the consent of the relatives when possible. We find very few relatives who make any objection, in fact, we have many requests from relatives to have this operation performed" (Clark 1924, 100).

This practice of obtaining "consent" seems quite common in the reports of other superintendents as well. From an ethical point of view, it was unlike what would today be considered informed consent; in some cases, it was explicitly conditional on the patient's mental health improving or to their release from the institutions.[15] Consent can probably be best understood as a form of legal insurance rather than a bow to patients' rights: by securing consent in some form, the hospital superintendents felt they were avoiding any possible legal trouble in the future. Ordering sterilization against a guardian's consent was challenged only once in California, in a case that was dismissed on a legal technicality.[16] Nevertheless, despite ethical deficiencies in this form of "consent," it does attenuate the image of the authoritarian state usually associated with compulsory sterilization. Braslow's work paints a rich picture of sterilization at Stockton as something done not in the spirit of diminishing patients, but enabling and even curing them.

Sonoma State Home

Sonoma State Home was California's chief institution for the mentally deficient and developmentally disabled, and was known until 1909 as the California Home for the Care and Training of Feeble-minded Children. It had by far the most vigorous sterilization program in the state from 1918 until 1949, under the long tenure of Dr. Fred O. Butler, perhaps the most eugenically inclined of all the California superintendents. A trained surgeon, Butler performed a number of the operations himself in the early days of his directorship; over 5,500 people were sterilized at Sonoma on his watch, and he maintained close contacts with California eugenics organizations throughout his career (Kline 2001, 81–98; Stern 2005, 106–107).

As recalled in an oral history interview many decades later, Butler set sterilization as one of his top priorities almost immediately after taking on the directorship after the death of his predecessor:

I proposed at our first board meeting, when asked what I had in mind to do for the improvement of mental retardation in California, I said that the first thing is to get plenty of water, and second, we should start the program of sterilization. . . . The board approved it providing I got the approval of other departments in Sacramento including the governor. I obtained their approval, and started within three months doing sterilizations, with the idea that we would get social service workers to help in planning and training, so by 1919 we were placing patients out, by giving them jobs, or letting them return home after they had been sterilized. (Butler 1970, 2)

For Butler, the ability to discharge patients was a high priority, both for economic reasons (state institutions were always felt to be overburdened and underfinanced) and because he expressed doubt that institutional life was positive for patients. In his mind, sterilization was one aspect of an overall approach that would allow patients to be successfully cared for in a noninstitutional context.

Butler believed that heredity played an important role in mental deficiency, but his belief in sterilization was multifaceted. Much of it was couched in the language of deinstitutionalization, of the importance of patients being able to "safely" return to the outside world; as at Stockton, the concern seems to have been more for patients' welfare than their liberty rights. In a 1921 letter, which Laughlin reprinted, Butler couched this self-sufficiency in explicitly hereditarian terms:

I think sterilization of a certain class of our inmates is most important; aside from the training and discipline obtained while here, the operation for sterilization renders them unable to propagate their kind; therefore, many of them are able to go on parole or be discharged and make their way in the world. This relieves the state and counties of the expense for their support as well as making them happy in the thought of being self-supporting. This procedure naturally makes more room in the institution for that class not able to cope with outside conditions, and relieves the relatives and various organizations of this burden. (Laughlin 1922, 59)

Yet Butler's belief in the importance of sterilization for self-sufficiency was not strictly limited to the idea that mental deficiency was spread through heredity. In 1930, at the height of his own eugenic interests, Butler observed that at his institution, "we have reached the point where we practically disregard whether they are the hereditary or non-hereditary type, for the reason that rarely is it possible for a feeble-minded mother to care for children properly" (Kline 2001, 100). This is a language he would return to in later comments on the subject as well: "Chiefly [sterilization] would give [the mentally disabled] an opportunity to have a normal life, extramurally without the burden of children. Of course I felt that no person should have children, even though [mental deficiency] might not be hereditary, [if] they are not able to care properly for children in their home. Therefore I felt that unless a person can be born of a normal parent, they'd better not be born at all. That was always my premise and I still carry that though, whether I am right or whether I am wrong" (Butler 1970, 4).

Butler's account of his activities decades later is largely unapologetic and no doubt highly self-serving, but aligns with many of his earlier writings as well. Even this die-hard eugenicist saw fit to argue that sterilization was, in his mind, a beneficial activity for patients even if heredity was a minor factor in their condition. It is a paternalistic attitude similar to Clark's rationale for the sterilization of mentally ill women: the "burden" of children would prevent marginal individuals from being fully self-sufficient. There is no doubt that Butler was a eugenicist—he was explicitly concerned with the dangers posed to society by mentally deficient people, believed in a hereditarian component, and was a frequent correspondent with Popenoe and other eugenicists (and, after retirement, became medical advisor for a voluntary sterilization group founded by Popenoe). Yet even Butler claimed to get more out of sterilization than eugenics alone, and he justified it to himself on grounds of benefit to the patient as well as to society. Even the most eugenic of the superintendents had multiple reasons to support sterilization, including his commitment as a physician to improving the health of individuals (Lombardo 2008, chap. 17).

Agnews State Hospital

Agnews State Hospital, an institution for the mentally ill located near San Jose, presents a foil to Stockton and Sonoma because sterilizations did not occur there in large numbers, despite its being an institution of considerable size and facing the same pressures as those that did sterilize. Unlike some other institutions that sterilized in middling to low numbers, Agnews was operating at full strength for most of the time during the years that sterilizations took place. Some places, such as Mendocino, had such a high level of superintendent turnover (nine different superintendents over the course of some forty years, and none there longer than eleven years) that their low sterilization rates quite possibly reflected simply a lack of coherent policy. Agnews, on the other hand, was under the control of a single administrator, Dr. Leonard Stocking, from 1903 to 1931. Why, then, did it have such a low rate of sterilizations?

It appears that Stocking himself had little to no enthusiasm for sterilization. This does not to appear to have been rooted in a strong belief in patient autonomy, but rather in his own idiosyncratic beliefs about the nature of mental health. His opinions on mental health are difficult to summarize, in part because they were originally vague and changed quite often: although he was not strictly a Freudian, by his own admission, through the 1920s he flirted with quasi-psychoanalytical approaches to mental illness, augmented by various physiologically inspired forms of treatment. In one report it was electrical stimulation, in another ultra-violet ray exposure, and in a yet later one hypnotic trance states—always described by Stocking as "new" treatments. When he did sterilize, it seems to have been for generally noneugenic reasons. In 1921, for example, Stocking requested the sterilization of a woman patient because "further pregnancies would be a decided hindrance for (the patient) remaining able when she again goes home," and he gave no indication that he considered heredity to play any role in her situation (Laughlin 1922, 54).

Stocking's views of mental health not only did not hold heredity as central, but also disparaged those who thought it was important. As he wrote in 1930, "Though heredity is of great importance in physiology, it is of only minor importance in psychology. Physical and psychical inheritance do not connect and run together" (Stocking 1930a, 221). In long essays on psychology published in a biennial report just a year before his death, Stocking mused that believing in heredity as the source of mental illness was like thinking that electric lighting was dependent on the presence of a switch: "But the novice in electric lighting is somewhat nonplussed when he notices that another light in the same room burns independent of any chain. An inhuman murderer comes from normal, respectable family, or a genius is born to commonplace, uninspired parents, and the heredity theory is somewhat discredited" (Stocking 1930b, 199).

In any case, Stocking did not sterilize in great numbers, which looks to have been a deliberate choice. Like Clark and Butler, he was given a free hand to act on his own idiosyncratic beliefs about the nature of mental health, and when he did sterilize it was for paternalistic reasons, though he too always claimed to get patient consent before sterilization. If he is a hero in this story, he is an odd hero, distinguished by ad hoc, changing beliefs about mental health, rapidly changing therapeutic strategies, and remarkable primarily for his choice not to sterilize, a choice not stemming from any apparent fundamental ethical regard for patient rights.

The foregoing evidence can be criticized as being rather broad, and concerned with superintendents' writings rather than records of daily practice (with the exception of Braslow's work, which is one of the few studies benefiting from access to such archives), and characterizes only three institutions with long-serving superintendents. Nevertheless, I think these cases quite vividly illustrate the relationship of institutional autonomy and sterilization practice, as well as the range of possible opinions that hospital administrators could hold toward sterilization. They also point toward what a more thorough study of sterilization in California would look like: considering hospitals as singular sites and paying close attention to what influences actually mattered in the case of individual, powerful superintendents, rather than gesturing vaguely toward connections between "popular" attitudes and sterilization practices.

Eugenics?

Sterilization and its motives in California were, as we have seen, deeply tied to the individual beliefs and personalities of the superintendents who ordered them. Those superintendents who had long tenures, such as Clark, Butler, and Stocking, exercised disproportionate influence over state sterilization trends compared with superintendents with very short tenures, not to mention disproportionate with respect to eugenics advocates, strategists, or think tanks. Parsing out the sterilizers' motives, and determining which of them deserve to be called "eugenics," requires some discussion.

As Diane Paul has noted, *eugenics* "is a word with nasty connotations but an indeterminate meaning. Indeed, it often reveals more about its user's attitudes than it does about the policies, practices, intentions, or consequences labeled" (Paul 1994, 143). Used almost exclusively in a pejorative sense in contemporary discourse, eugenics has been applied to practices from expectant mothers taking vitamins for the health of the fetus to the worst atrocities of Nazi genocide (Mahowald 2003). As Elof Axel Carlson aptly put it:

Historians of science, however, have found that the term is chameleon-like, changing definition, purpose, scope, and values in different eras, countries, and social settings. At one extreme, eugenics is a gigantic umbrella that covers almost all social movements in which sex, gender, heredity, family planning, reproductive options, marriage, immigration, social status, and social failure are involved. It ranges from concerns about the most dependent children and adults to interest in the most successful and eminent high achievers and their roles in shaping future generations of humanity. At a more restricted level of historical interpretation, eugenics is the application of human heredity to an analysis of differential birthrates. The broader historical approach makes eugenics a more difficult target for those concerned about personal liberties. The narrower approach makes the old-line eugenics of the first half of the twentieth century a dead horse that is no cause for present worry. (Carlson 2003, 761)

Taking the "umbrella" view, to use Carlson's term, has historical justifications: the eugenics *movement* was a disparate group of individuals with beliefs that shared family resemblances but were not at all necessarily connected. Many eugenicists had very low regard for one other and routinely disagreed with others in the community. No one has illustrated these historical discontinuities better than Diane Paul, who herself has argued that the attempt to stipulate a definition of "eugenics" is usually a meaningless exercise that avoids discussion of politics and ethics (Paul 1994, 1995, 2007).

At the same time, when dealing with historical episodes, as opposed to determining future policy, there can be value in making distinctions. If eugenics can encompass everything from vitamin consumption to genocide, then it loses its meaning. As my purpose is not to establish "guilt by association," but rather to understand the motives of the California superintendents and the means by which they were enabled by their institutions, I would argue for the relatively limited definition of eugenics proposed by Braslow. That definition refers to policies aimed at effecting *collective* change—at the level of society, nation, the human race, or the gene pool—rather than the *individual*. As Paul has noted, this is not an uncommon demarcation criterion, and is often favored by those who argue that technologies of consumer-driven genomics are *not* "eugenics" (Paul 1994, 144–149).

One could potentially abandon the term "eugenics" altogether, and refer only to *hereditarian* concerns, as it is exactly the hereditarian concerns, coupled with state coercion, that disturb people most about the eugenics of the first half of the twentieth century. Sexism, racism, and even authoritarian tendencies were often coupled with and mutually reinforced by hereditarian eugenics in the early twentieth century, but they were not necessarily an integral part of that worldview. To consider racism a central component of eugenics simply because the two were often coupled is similar to considering racism a central component of Darwinian evolution simply because many proponents of Darwinism held what would be today considered racist beliefs. For purposes of this volume it is important to recognize that eugenics did not inevitably lend scientific support to other pernicious social ideologies: one could be a eugenicist without being a racist, and a sexist without being a eugenicist, and so forth.

In the final analysis, though, attempts to fit practices into preformed and pejorative categories miss a larger point. As Johanna Schoen has argued, reproductive regulations existed on an ethical "continuum"—at times enabling, at other times destroying, reproductive autonomy (Schoen 2005, 7). Understanding the implementation of these practices, and focusing on the specific institutional instantiations and translations of various ideological and public health goals, must take precedence over a simple classification of practices into those that resemble Nazis and those that do not.

The California superintendents surveyed thus far clearly sought a number of different goals through sterilization. The designation *therapeutic* sterilization is best reserved for those who hoped, like Stockton's Clark, that the procedure would effect marked physical and mental benefits for the patient. Sterilizations ordered because physicians believed that childbirth would cause another breakdown could be considered *preventive*, whereas those motivated by the belief that the patient was mentally unfit to be a good parent could be called *paternalistic*. In this case, the paternalistic impulse arose from the physician's conviction that he or she was better qualified than the patient to judge the patient's fitness as a parent.[17] And finally, of course, there was the purely *eugenic* or *hereditarian* justification, which I have defined as any intent to reduce the incidence of mental illness or mental deficiency in society at large by blocking the transmission of "defective" genetic material.

All of these justifications seem to have been more or less sanctioned by the state Department of Institutions in approving sterilization requests from the superintendents. The amount of oversight appears to have been quite low—short letters from superintendents requesting sterilization were all that was required, and there are cases of patient diagnoses being specially modified to make sterilization easier.[18] In many cases, multiple motives were obviously at play. Sonoma's Butler, as we have seen, sterilized mainly for eugenic and paternalistic reasons, and Stockton's Clark sterilized for therapeutic, paternalistic, and preventive reasons. None of these rationales were mutually incompatible, and in the end eugenics advocates such as Popenoe and Gosney could be reasonably happy with sterilizations for any reason as long as they were taking place within mental hospitals. Only the case of Stocking at Agnews points to a specific intent *not* to sterilize, but he was the exception, and even he approved at least some paternalistic sterilizations, though he rejected eugenics.

The Consequences of Decentralization

The sterilization trends in California can be traced, as we have seen, largely to a handful of superintendents who either had strong reasons to sterilize or strong reasons not to. Institutions that had individual superintendents for long tenures (i.e., multiple decades) tended to have extremely high sterilization rates, with the exception of Agnews State Hospital, whose superintendent was simply not interested in sterilization. Institutions with rapid turnover in superintendents all have middling rates. The delegation of discretion to superintendents resulted in many sterilizations clustered in few institutions, and carried out in some cases with documented non-eugenic motivations.

Despite its somewhat confusing language, the law enabled superintendents to implement their policies efficiently. By the 1940s, they even had standardized forms for requesting sterilization, with tiny checkboxes to indicate various patient afflictions, and even a section dedicated to consent from legal guardians—all so routinized that requests could be approved the very day they were received (figure 2.3). Perhaps the plurality of justifications supported by California policy also aided in maintaining the superintendents' authority: sterilization could be, as the State Attorney General had argued in 1910, just another medical procedure for physicians to use in accordance with their professional judgment (Webb 1910). By law, the California superintendents could sterilize, and because of the multiple reasons that the law recognized and the medical culture of the period sanctioned, they usually did. Importantly, the law had no mechanisms for appeal.

California's legal situation allowed its superintendents to start sterilizing earlier than in most institutions in the country, and the flexibility of the law drove California to become the premier state for sterilization. Although California superintendents felt free to request sterilizations for

Operation by Dr. Traver 7.4.45

Recommendation and Approval for Vasectomy & Salpingectomy
for the Purpose of Sterilization

FORM 787

5

Name_____

Institution __PATTON STATE HOSPITAL__

RECEIVED

JUN 18 1945

DEPT. OF
INSTITUTIONS

PERSONAL HISTORY

HOSPITAL CASE No._____

Age__27 yrs.__Nativity__California__Religion__Protestant__Education__High School__

Marital status__married__Sex__Female__No. of children__One__Ages__Seven yrs.__

FAMILY HISTORY (for additional space use reverse side): None

CLINICAL HISTORY (for additional space use reverse side):

Date admitted__May__1945__Present diagnosis__Dementia Praecox, Catatonic Type.__

Attacks (previous, and diagnosis of each): Norwalk St.Hosp. 5-22-42 to 3-24-43 Manic Depressive
Psychosis, Depressed Type.

LEGAL PROVISIONS (compliance with):

This form is submitted in accordance with section 6624 of the Welfare and Institutions Code of the
State of California.

1. Legally signed and prepared commitment papers { are / ~~are not~~ } on file at this institution.

2. This patient is afflicted with:

   [X] Mental disease which may have been inherited and is likely to be transmitted to descendants.

   [ ] Feeble-mindedness, in any of its various grades (specify grade) _____

   [ ] Perversion or marked departures from normal mentality.

   [ ] Disease of a syphilitic nature.

Written consent { ~~not given~~ / given } by_____   __Mother__
                                              Name   Relationship

under date of__June 4, 1945__copy of which is attached hereto.
                (If consent not given, submit separate letter giving circumstances)

After careful consideration of the case of_____
by the members of the Medical Staff of this institution, it is their belief that this patient is suffering
from the affliction above noted and it is their recommendation that the operation for the purpose of
sterilization be performed, with which opinion and recommendation I concur and do hereby request
your approval.

[DATE]__June 14,__1945

G. W. Webster
Medical Director and Superintendent

Approved and authorization for an operation for sterilization granted this__18th__day

of__June__1945.

Dora Shaw Heffner
Director of Institutions

Figure 2.3
Form 787, "Recommendation and Approval for Vasectomy or Salpingectomy for the Purpose
of Sterilization," submitted to the California Department of Institutions June 1945, by G. M.
Webster, medical superintendent of Patton State Hospital, and approved by Dora Shaw
Heffner, Director of Institutions. *Source*: Department of Mental Hygiene, Mendocino State
Hospital Records, California State Archives, Sacramento, California. Nongeneral personal
information has been blacked out by author.

varied reasons, physicians elsewhere complained that their laws were cumbersome, contradictory, and subject to legal challenges (Laughlin 1922, 52–92). This institutional explanation for why California's sterilization rates were so much higher than in any other state is appealing in that it requires no abstract notions such as public opinion, state or national fads, or even mainstream scientific or medical opinion. Rather, the impetus to sterilize, in this narrative, falls squarely on the shoulders of the individuals who had the means, motives, and opportunity to sterilize.

Between 1951 and 1952 the rate of sterilizations dropped by 80 percent, and from then on the practice declined to less than a half-dozen per year in 1960 (figure 2.4). This abrupt change came with no fanfare and no hand-wringing, no comparisons to Nazi Germany, and no discussion of rights to reproduction. The horror we attach to the sterilizations today, and to eugenics in general, did not become widespread until the 1970s, with the rise of interest in patient autonomy, women's rights, the power of the medicalized state, and a right to reproduction that were conspicuously absent from earlier discussions of eugenics (Paul 1995, chap. 7; Lombardo 2008, chaps. 17–19; Paul 2002). In other words, profound social transformations happened in American understandings of the human body and its rights in the late twentieth century, but they do not seem to account for the rise and fall of sterilization.

So why did California stop sterilizing when it did? A number of organizational changes seem important. First, the newer superintendents did not sterilize at the rates of the older ones. This is most likely due to changing medical attitudes toward mental health. The last of the enthusiastic sterilizers, Fred O. Butler, recalled the change:

Oh, [the mindset] changed materially. Well it was shortly after I left up there. I know I went back about the following year or two, the superintendent [Dr. Porter] asked me to come back and talk on it, on sterilization, and I found the dissenters on it were mostly psychologists. They didn't agree, and social workers were second, and physicians, I think, were third, I would say. That is, [they] question[ed] the advisability and so forth of sterilization. Some thought that we didn't have enough information as a basis for sterilization. Of course they didn't go for my premise that they should be sterilized regardless of their heredity or not. (Butler 1970, 14)

Butler himself retired in 1949, having served as the director of the Sonoma State Home for over three decades.[19] At the time, Sonoma was responsible for 53 percent of all annual sterilizations in California—four times as many as any other hospital. With Butler gone, there were no strong advocates for sterilization left in the California system.

At the same time that enthusiasm for sterilization was winding down within the medical system, important organizational shakeups were

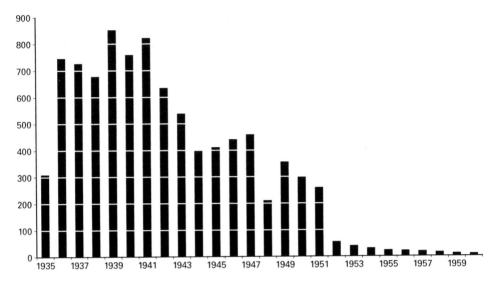

Figure 2.4
Annual sterilizations, all California state hospitals, 1935–1960. *Source*: Data from "Sterilization Operations in California State Hospitals for the Mentally Ill and for the Mentally Deficient, April 26, 1909 through June 30, 1960," Statistical Research Bureau, California State Department of Mental Hygiene (October 5, 1960), received from the California Department of Mental Health (Statistics and Data Analysis), on July 17, 2003. Graph by author.

occurring in the state. In the late 1940s, Governor Earl Warren developed a strong personal interest in modernizing the California mental health system. He considered the old institutions (like Stockton, Agnews, and Sonoma) to be "regular 'snake pits,'" a throwback to the previous century, so "appalling" that after visiting a number of them, "I did not have a peaceful night's sleep for over a month." Moreover, the hospitals were, he later recalled, "a loose organization that left a great deal of local autonomy and enabled each group [of administrators] to operate more or less as a political entity." He wanted comprehensive reform—literally destroying the nineteenth-century brick buildings that were still housing some patients—in order to "take California out of the asylum age and put it in the hospital age" (Warren 1977, 178–183).

The same year that Butler retired from Sonoma, 1949, Warren appointed a new director to head up the overall organization, which he had renamed the Department of Mental Hygiene. Dr. Frank F. Tallman was an outsider to the California system, brought in from an identical position in Ohio (a state that never had sterilization legislation). Neither Tallmar nor Warren ever publicly distanced themselves from the state sterilization program. Both were interested in general organizational reform. Tallman saw as one of his chief objectives the centralization of the Department of Mental Hygiene's power over its sprawling institutions, full of superintendents whom he derided as "foot-draggers" (Tallman 1973, 42). He later recalled that in this, he had a partnership with Warren: "[Warren] felt that the superintendents of the individual institutions should have a less parochial and isolated attitude and role and ought to be helped to be active in community mental health affairs. Also he thought that they needed encouragement in realizing the helpful importance of the central governing body—namely the State Department of Mental Hygiene" (Tallman 1973, 19–20).

To implement this policy, Tallman initiated a rewrite of many state mental health statutes, including the sterilization law. The 1951 revision, which modernized the statutory language and made it more compliant with *Buck v. Bell*—adding an appeals mechanism, for example—changed little in the procedures for requesting patient sterilization.[20] Though the law changed very little on its face, the attitude of those responsible for enforcing it shifted quite a lot. They were, privately, dubious of the local expertise of the hospital superintendents, and wanted far more documentation of consent and need than before.[21] Sterilization was now a centralized concern, and the few official statements on sterilization available from his department explicitly endorsed what I have labeled paternalistic sterilization and omitted reference to any hereditarian motivation.[22]

Following the centralization of state mental health policies, and a tightening up of the state law, sterilization rates plummeted. What ended sterilization in the state was not a reframing of eugenics as an ethically unacceptable practice endangering basic human rights, nor any apparent association of the practice with National Socialism. No one took credit for killing the practice, and no one at the time appears to have noticed that it had ended. Administrators interested in bureaucratic efficiency centralized the system and reduced the autonomy of hospital superintendents at about the same time that the last influential advocate of sterilization retired. Sterilization in California died not with a bang but a whimper.

If the new mental hygiene administrators seemed unconcerned with the practice of sterilization, neither was the public. The *Sacramento Bee*, which devoted ample attention to any state legislation of perceived significance, put the notice of the 1951 revisions to the sterilization law in a single line of text, in the smallest possible font, typically used for only the most uninteresting legislation: "SB 730, Dillinger—Eliminating sex perversion of syphilitic disease as the basis for sterilization of persons in state mental institutions and allowing sterilization for mental illness or mental deficiency only. Also, provides for notices and hearings on intended sterilization."[23] The *Los Angeles Times* ran a similarly brief notice.[24]

Behind-the-scenes evidence indicates that modernizing administrators such as Tallman found the existing sterilization practices out of date, not ethically disturbing. In a 1951 memo to Warren, Tallman emphasized the parts of the new law that centralized the ultimate discretion for sterilization with the Director of Mental Hygiene (that is, himself), and improved patient and family notification. He also pointed out that they had carefully crafted the new law to avoid offending the Roman Catholic Church, the only long-time opponent of sterilization.[25] This and other correspondence about the bill makes it clear that the state administrators were starting to see sterilization as something to be sparingly used and not of primary importance to the overall treatment of mental health.

California's sterilization boom was institutional in its beginnings and institutional in its end. Broader "scientific ideas" were not irrelevant—indeed, part of my explanation in this chapter rests on changing conceptions of mental illness, which meant that new administrators coming into the system were less focused on heredity than previous generations. But the end of sterilization did not come about simply because ideas changed. Rather, it came about because of specific changes in institutional and legal structures for managing public health. That attitudes towards sterilization changed is an insufficient explanation by itself, without taking into account the way that discretionary authority was distributed throughout the mental health bureaucracy.

A practice-based account of sterilization in California emphasizes that the link between ideology and action is rarely direct, but instead is mediated by specific legal, social, professional, and organizational factors. In California, the policies of relatively few individuals with long tenures had massive statistical effects in a system that delegated power over the bodies of mental patients to the institutions that held them. This decentralization of power—or perhaps more accurately its distribution into independent nodes—is a familiar feature of American bureaucracy.

We might in conclusion extend, briefly, this analytic approach to the infamous sterilization program in Germany. Like so many other efforts at "coordination" (*Gleichschaltung*) in the German state, the German sterilization law attempted to eliminate discretion among practitioners, making mandatory the reporting of patients to be sterilized. Under German law, physicians—not merely those working in state institutions—could be fined for not reporting candidates for sterilization to centralized sterilization courts.[26] The hierarchical Nazi system made a conscious effort to ensure that ideology was intimately tied to law and practice, extending state authority into the capillaries of bureaucratic behavior; in the United States as a whole, and especially in California, there was no such attempt at top-down "coordination."[27] This structural difference, combined with the fact that the Nazi sterilization law applied to the general population (not simply those in state institutions), no doubt accounted in large part for the ability of the German regime to sterilize over four hundred thousand individuals, most in the first four years of operation; six decades of sterilization laws in the United States did not reach one-sixth that figure (Proctor 1988, 108; Bock 2004).

The history of compulsory sterilization in California holds a complicated lesson for commentators on genetics in society. The results, in this case, were less about sweeping, science-driven ideas about individual and social health than they were about the idiosyncrasies of an enabling system; they were less about the overall coordination of a grand plan than they were about unchecked local authority and discretion. Ethically, that makes the issue of what precautions should be taken in the future more problematic in a sense: instead of being motivated by a single, wrong-headed idea of genetic perfection, which could be fought with "better" science, we see an unfortunate convergence of heterogeneous motives, most centered around faulty perceptions about what would be most beneficial to the individual patient, not to society as a whole. We do not find Nazis in Californian mental health institutions.

Tying this chapter to the broader cultural and normative concerns of this volume, we find Americans doing things in an American way, yet, when judged by the sheer number of sterilizations, ending up with results disconcertingly similar to those attained by German centralized coordination. One is tempted to speculate that any future eugenics in America, however defined, will remain of a distinctly American character: decentralized, well-intentioned, quiet, but—if left unregulated—deeply troubling in its ramifications.

Acknowledgments

The number of people who have read, heard, and given opinions on various iterations on this long-evolving work are too many to list at this point. For this version in particular, and for the encouragement to publish it, I must acknowledge the help of Ellen Bales, Alison Bashford, Cathryn Carson, Miroslava Chavez-Garcia, Sheila Jasanoff, Diane Paul, and the anonymous reviewers.

Notes

1. American sterilization statistics are often quite difficult to come by in consistent datasets (that is, those made with the same methodologies, counting the same things). Most sterilization statistics are self-reported and compiled by eugenics organizations (with various levels of detail and accuracy). The most complete single dataset for national sterilization statistics is Robitscher 1973, appendix 2.

2. The first significant postwar history of eugenics was Haller 1963. Previous "histories" of eugenics were primarily works by active eugenicists, for example, Landman 1932. The current standard account of eugenics in the United States has long been Kevles [1985] 1995. Since the Human Genome Project, new monographs on American eugenics have been appearing at the rate of at least one per year.

3. An important exception to this is the recent book by Alexandra Minna Stern, which goes to considerable lengths to understand California eugenic thought as a phenomenon in and of itself. Stern 2005, especially chap. 3. Similarly, Edward Larson's book on eugenics in the South does an excellent job of providing a re-gionalized analysis that combines the more traditional intellectual approach with the specifics of the southern legal and political context (Larson 1996).

4. There is something of a deliberate muddling here between the two uses of the terms "institution": the specific sense of state institutions for the mentally ill (hence California's Department of Institutions), and the more general sense of institutions of government, power, and society. In this particular case, there is a considerable overlap between the two meanings. In the latter sense of the meaning, particularly useful is Lenoir 1997.

5. There were two additional amendments to the sterilization laws in 1923, but they pertained to special cases and did not change the overall framework of the legislation.

6. California Statutes of 1909, Chapter 720, 1093.

7. That castration was not considered to be a valid use of the law was made clear early on by the state Attorney General, who strongly favored the use of vasectomy for males (Webb 1910). Compare "Casts Doubt on New State Law: Validity of an Asexualization Act is Questioned by the Attorney-General," *San Francisco Chronicle* (March 6, 1910), 29. Early sterilizations were in fact done by castra-tion; vasectomies were still a new procedure, having only been developed in the 1890s (Gugliotta 1998).

8. Frank F. Tallman to Earl Warren, Inter-Departmental Communication, "Sub-ject: Assembly Bill 2683" (March 31, 1953), in the California State Archives, Sacramento, California, file on Assembly Bill 2683. This was in the context of a legal review of the statute completed after the practice had essentially stopped.

9. California Statutes of 1913, Chapter 363, 775.

10. California Statutes of 1917, Chapter 489, 571; California Statutes of 1923, Chapter 224, 448. "Punitive" sterilization—that is, sterilization as a punishment for prisoners, was rare in the California context, as it was considered from the beginning to be likely unconstitutional, a point reaffirmed years later by the Su-preme Court in *Skinner v. Oklahoma* (1942).

11. California Statutes of 1951, Chapter 552, 1706.

12. All of these institutions went through at least one name change over the years. The names used here were what they were referred to as during the majority of the period here surveyed. DeWitt and Modesto State Hospital were, during this period, primarily for the mentally ill, but also were equipped for a small amount of the mentally deficient as well, hence their dual-use designations in figure 2.2.

13. Earlier work on the "revitalizing" effect of testicular fluids was performed by the nineteenth-century physiologist Charles-Édouard Brown-Séquard, whom Clark credited highly (Braslow 1996, 39–40).

14. Braslow suggests that the therapeutic motivation may have been responsible for sterilizations in California outside of Stockton as well; the *Biennial Reports*, however, seem to indicate that while other superintendents elsewhere were aware of Clark's claims, they is little evidence that they subscribed to them. There is evidence, however, that therapeutic goals played a role in some sterilizations in other states, as well (Gugliotta 1998).

15. Paul Lombardo makes a strong case for how flimsy "consent" claims could be in other states—that a physician could easily mislead a patient as to the permanence and nature of the operation (Lombardo 2008, 247–248). For further complications on the question of voluntary/involuntary sterilizations, see especially Schoen 2005.

16. *Garcia v. State Department. of Institutions,* 36 C.A. 2d, 152.

17. The statutes also permitted *punitive* sterilization as a form of punishment for a crime. Punitive sterilization played a larger role in other state programs before it was declared unconstitutional in *Skinner v. Oklahoma* (1942), but was never widely used in California, where it was judged early on to be constitutionally problematic by the state Attorney General (Webb 1910). On punitive sterilization in Oregon, see Largent 2002; in Indiana, see Gugliotta 1998.

18. Braslow gives evidence of an instance in which physicians at Stockton agreed upon an official patient diagnosis, which would make it easier to request a (therapeutic) sterilization (Braslow 1996, 37).

19. Butler then became the medical advisor for a *voluntary* sterilization group founded by eugenicist Paul Popenoe.

20. California Statutes of 1951, Chapter 552, 1706. On the purpose of the law, see Wallace G. Colthurst, Deputy Attorney General, to Earl Warren, Inter-Departmental Communication, "Subject: Senate Bill No. 730" (May 14, 1951), and Frank F. Tallman to Earl Warren (May 15, 1951), both in California State Archives, Sacramento, CA, file on S.B. 730/A.B. 2037.

21. An interesting but problematically retrospective look at various concerns that some hospital administrators had about sterilization can be found in Tarjan 1998, 204–245.

22. See, for example, the short blurb in the *Biennial Report* for 1950–1952, on 63, which begins by acknowledging that those "interested in the study of eugenics" had asked questions about the sterilization policy, but then immediately emphasized that "the ultimate therapeutic benefit to the patient is the chief concern of the

medical staff of each hospital," by which the threat of pregnancy (not "therapy" in the sense of Stockton's Clark) was what was to be averted.

23. *Sacramento Bee* (May 23, 1951), 12. To give a sense of context, it is worth noting that the line is placed under a five-paragraph article on the governor vetoing a bill which would give thirty days pay to state workers who joined military service, and over an article announcing that, "Congressmen may visit Sacramento in redistrict study." The line is listed with another bill which was signed at the same time by the governor which gave the Department of Mental Hygiene the power to enforce regulations on private homes for the mentally ill.

24. Associated Press, "The Day in Sacramento," *Los Angeles Times* (May 23, 1951), 7. Of note is that on the same day the *Los Angeles Times* ran its own one-line summary of the revision of the state sterilization law, it carried a far more prominent headline relating to a city Supervisor who declared that "sex criminals be incarcerated, sterilized or thoroughly controlled." "Sterilization Urged for Sex Degenerates," *Los Angeles Times* (May 23, 1951), 4.

25. Frank F. Tallman to Earl Warren (May 15, 1951), California State Archives, Sacramento, CA, file on S.B. 730/A.B. 2037.

26. "The law for the prevention of hereditarily diseased offspring (Approved translation of the '*Gesetz zur Verhütung erbkranken Nachwuchses*'). Enacted on July 14, 1933. Published by *Reichsausschuss für Volksgesundheitsdienst*" (Berlin: Reichsdruckerei, 1935), 12 (Order 1, Article 9). The fine was up to 150 Reichmarks per patient (around $750 in current dollar value).

27. On the National Socialist "coordination" of the medical profession, see especially Proctor (1988). It is of note that even with such "coordination," there was plenty of room for power struggle between rival factions (e.g., Walker 1995, chaps. 2–3).

References

Black, Edwin. 2003. *War against the weak: Eugenics and America's campaign to create a master race.* New York: Four Walls Eight Windows.

Bock, Gisela. 2004. Nazi sterilization and reproductive policies. In *Deadly medicine: Creating the master race*, ed. Dieter Kuntz, 61–88. Washington, D.C.: United States Holocaust Memorial Museum.

Braslow, Joel T. 1996. In the name of therapeutics: The practice of sterilization in a California state hospital. *Journal of the History of Medicine and Allied Sciences* 51 (1): 29–51.

Braslow, Joel I. 1997. *Mental ills and bodily cures: Psychiatric treatment in the first half of the twentieth century.* Berkeley: University of California Press.

Butler, Fred O., interviewed by Margot W. Smith. 1970. *Interview on mental health and mental retardation* (rough transcript). Berkeley: Regional Oral History Office, Bancroft Library, University of California.

California Department of Mental Hygiene. 1950. *Statistical Report.* Sacramento: California State Printing Office.

Carlson, Elof Axel. 2001. *The unfit: A history of a bad idea.* Cold Spring Harbor, N.Y.: Cold Spring Harbor Laboratory Press.

Carlson, Elof Axel. 2003. Review of Wendy Kline, *Building a better race: Gender, sexuality, and eugenics from the turn of the century to the baby boom. Isis* 94 (4): 761.

Clark, Fred P. 1922. Report of the medical superintendent of the Stockton State Hospital. In *Biennial report,* California Department of Institutions, 88. Sacramento: California State Printing Office.

Clark, Fred P. 1924. Report of the medical superintendent of the Stockton State Hospital. In *Biennial report,* California Department of Institutions, 100–103. Sacramento: California State Printing Office.

Gosney, E. S., and Paul Popenoe. 1929. *Sterilization for human betterment: A summary of results of 6,000 operations in California, 1909–1929.* New York: Macmillan.

Gugliotta, Angela. 1998. "Dr. Sharp with his little knife": Therapeutic and punitive origins of eugenic vasectomy—Indiana, 1892–1921. *Journal of the History of Medicine and Allied Sciences* 53 (4): 371–406.

Haller, Mark H. 1963. *Eugenics: Hereditarian attitudes in American thought.* New Brunswick, N.J.: Rutgers University Press.

Kevles, Daniel J. [1985] 1995. *In the name of eugenics: Genetics and the uses of human heredity.* Cambridge, Mass.: Harvard University Press.

Kline, Wendy. 2001. *Building a better race: Gender, sexuality, and eugenics from the turn of the century to the baby boom.* Berkeley: University of California Press.

Kühl, Stefan. 1994. *The Nazi connection: Eugenics, American racism, and German National Socialism.* New York: Oxford University Press.

Landman, Jacob H. 1932. *Human sterilization: The history of the sexual sterilization movement.* New York: Macmillan.

Largent, Mark A. 2002. The greatest curse of the race: Eugenic sterilization in Oregon, 1909–1983. *Oregon Historical Quarterly* 103 (2): 188–209.

Larson, Edward J. 1996. *Sex, race, and science: Eugenics in the deep South.* Baltimore: Johns Hopkins University Press.

Laughlin, Harry H. 1922. *Eugenical sterilization in the United States.* Chicago, IL: Psychopathic Laboratory of the Municipal Court of Chicago.

Lenoir, Timothy. 1997. *Instituting science: The cultural production of scientific disciplines.* Stanford: Stanford University Press.

Lombardo, Paul A. 1985. Three generations, no imbeciles: New light on *Buck v. Bell. New York University Law Review* 60 (1): 50–62.

Lombardo, Paul A. 2008. *Three generations, no imbeciles: Eugenics, the Supreme Court, and Buck v. Bell.* Baltimore: Johns Hopkins University Press.

Mahowald, Mary B. 2003. Aren't we all eugenicists? Commentary on Paul Lombardo's "Taking Eugenics Seriously." *Florida State University Law Review* 30 (219): 219–235.

Paul, Diane B. 1994. Eugenic anxieties, social realities, and political choices. In *Are genes us? The social consequences of the new genetics*, ed. Carl F. Cranor, 142–154. New Brunswick, N.J.: Rutgers University Press.

Paul, Diane B. 1995. *Controlling human heredity: 1865 to the present*. Atlantic Highlands, N.J.: Humanities Press.

Paul, Diane B. 2002. From reproductive responsibility to reproductive autonomy. In *Mutating concepts, evolving disciplines: Genetics, medicine and society*, ed. Lisa S. Parker and Rachel A. Ankeny, 87–105. Boston: Kluwer Academic Publishers.

Paul, Diane B. 2007. On drawing lessons from the history of eugenics. In *Reprogenetics: Law, policy, and ethical issues*, ed. Lori P. Knowles and Gregory E. Kaebnick, 3–19. Baltimore: Johns Hopkins University Press.

Popenoe, Paul. 1934. The progress of eugenic sterilization. *Journal of Heredity* 25 (1): 19–26.

Proctor, Robert. 1988. *Racial hygiene: Medicine under the Nazis*. Cambridge, Mass.: Harvard University Press.

Robitscher, Jonas B., ed. 1973. *Eugenic sterilization*. Springfield, Ill.: Thomas.

Schoen, Johanna. 2005. *Choice and coercion: Birth control, sterilization, and abortion in public health and welfare*. Chapel Hill: University of North Carolina Press.

Stern, Alexandra. 2005. *Eugenic nation: Faults and frontiers of better breeding in modern America*. Berkeley: University of California Press.

Stocking, Leonard. 1930a. The triune mind. In *Biennial report*, California Department of Institutions, 216–223. Sacramento: California State Printing Office.

Stocking, Leonard. 1930b. Balance. In *Biennial report*, California Department of Institutions, 199–209. Sacramento: California State Printing Office.

Tallman, Frank Ford, interviewed by Gabrielle Morris. 1973. *Dynamics of change in state mental institutions*. Berkeley: Regional Oral History Office, Bancroft Library, University of California.

Tarjan, George, interviewed by Michael S. Balter. 1998. *UCLA Neuropsychiatric Institute and Hospital oral history transcript, 1997–1998*. Los Angeles: Oral History Program, University of California, Los Angeles.

Walker, Mark. 1995. *Nazi science: Myth, truth, and the German atomic bomb*. New York: Plenum Press.

Warren, Earl. 1977. *The memoirs of Earl Warren*. Garden City, N.Y.: Doubleday.

Watson, James D. 2001. Genes and politics. In *A passion for DNA: Genes, genomes, and society*, 183–212. Woodsbury, N.Y.: Cold Spring Harbor Laboratory Press.

Webb, Ulysses S. 1910. "Official opinion of the attorney-general of California on the asexualization law" (March 2, 1910), 324–328. Reproduced in Laughlin 1922.

3

Making the Facts of Life

Sheila Jasanoff

On August 9, 2001, nine months after taking office and one month before the terrorist attacks that changed the course of his administration, U.S. president George W. Bush held his first nationally televised news conference. The subject was not Osama Bin Laden or Al Qaeda, news of which had already percolated into America's intelligence services, but a surprisingly partisan issue on the frontiers of biomedical science. The topic was research with human embryonic stem cells (hESCs). That August evening, from a monthlong working holiday at his Texas ranch, the president announced that he would permit federal funds to be used only for research on some sixty embryonic stem cell lines that existed as of that date; no newer cell lines would be covered. This policy, Bush said, would allow U.S. scientists "to explore the promise and potential of stem cell research without crossing a fundamental moral line, by providing taxpayer funding that would sanction or encourage further destruction of human embryos that have at least the potential for life" (Bush 2001). The theme that one form of "life" should not be sacrificed for the sake of others resounded throughout his presidency. For example, in July 2006 the president exercised his first legislative veto on H.R. 810, the "Stem Cell Research and Enhancement Act," while expressing "the hope that we may one day enjoy the potential benefits of embryonic stem cells without destroying human life" (Bush 2006).

Bush's statements about stem cells underscore the point made a generation or so earlier by the French social theorist Michel Foucault (1998) that life itself has become the primary object of modern governmental power; in governing life, states engage in activities that Foucault termed *biopolitics*. Since the latter part of the twentieth century, biology and biotechnology have transformed the territory of biopolitics. Public policy today concerns itself not only with governing potentially unruly human subjects, as in California's sterilization-happy mental health clinics (Wellerstein, chapter

2, this volume), but with a wide variety of constructs derived from nature and possessing indeterminate physical properties and moral valences. Laboratories and clinics teem with ambiguous entities: genetically modified plants and animals, immortal cell lines derived from mortal human bodies (Skloot 2010), embryonic and nonembryonic stem cells, induced pluripotent cells with the possibility of developing into full-blown life, human-animal chimeras, and engineered biomolecules with no natural counterparts but capable of self-replication. Associated with these objects are an increasing variety of new techniques that involve testing, removing, replacing, and recombining genetic material from organisms at varied stages of development.

Political leaders in industrial nations celebrate this productivity, seeing there the promise of cures for individual and collective ills: for individual bodies, freedom from grave yet incurable afflictions such as juvenile diabetes, cancer, spinal injuries, Alzheimer's disease, and Parkinson's disease; for aging and declining populations, an economic and social rebirth through technological inventions, wealth-creating jobs, and enhanced possibilities of living well in the world. Beneath these powerful sociotechnical imaginaries (Jasanoff and Kim 2009), however, anxiety bubbles. Is the scramble for reengineering life also instrumentalizing it? Do biologists' ingenious constructions commodify life and violate human integrity? Who is in charge of choice and change? Is anyone responsible for remaking the facts of life?

Around the world today, there is growing recognition that classical forms of politics, clustering around elections, are not sufficient to ensure democratic control over the answers to such questions (Callon, Lascoumes, and Barthe 2009; Gottweis 1998; Irwin 2001; Irwin and Michael 2003). Nor, as the long-drawn American abortion controversy illustrates, are people always satisfied with decisions that presume to settle basic bioconstitutional debates at the level of Supreme Court jurisprudence (Post and Siegel 2007). One indicator of unrest is the rise of bioethics not only as a field of expertise but also as an adjunct to public policy. In the quarter-century since Britain's Warnock Commission produced its seminal report on embryo research (Warnock 1985), bioethics has ceased to be merely a "fashion"—as one European Commission official described it to me in the quiet 1990s, before the cloning of the sheep, Dolly, in Edinburgh's Roslin Institute (Wilmut et al. 1997). Most Western governments, and increasingly developing states, have supplemented funding for the life sciences and technologies with public support for ethical analysis. Bioethics has come into its own as a new deliberative discourse and new questions have arisen about what this means for scientific accountability, democratic deliberation (Hurlbut 2010), and life itself.

In this chapter, I focus on how Britain, Germany, and the United States have confronted issues of responsibility for novel, boundary-crossing biological entities. Reinforcing this volume's coproductionist theme, I show that bioethical deliberation in each country serves as a site of ontological surgery—that is, for deciding how to describe and characterize the problematic entities whose natures must be fixed as a prelude to ethical analysis. "What *is* this thing?" Answers are needed before we can decide what to do with it. A quick example will illustrate the point. At a meeting of a bioethics committee I serve on, a member asked: "So, let me be sure I understand—if you take a rabbit egg and replace the nucleus with the nucleus from a human somatic cell, what you get is a human embryo, right?" Talk stopped until an inarticulate murmur ratified the speaker's speculation. It was a comforting consensus because we *know* what to do with human embryos, at least to some degree, even if it was not always thus (Hurlbut 2010). But the crucial point is that this was a defining moment for both nature and ethics. Almost unnoticed, biological and ethical confusion were tamed together to produce a way of moving forward in uncharted waters.

That dual work of biological classification and moral clarification proceeds, as I also argue in this chapter, in conformity with deeper scripts of acceptable deliberation in modern political cultures. Ontological surgery is part and parcel of a broader process of ontological politics (Jasanoff 2005a; Mol 1999), which in turn plays out within a nation's established bioconstitutional order. Drawing on my earlier research on the comparative politics of biotechnology, I first articulate some of that work's latent implications for national practices of ontological boundary-drawing, focusing on the intersection of classification, ethics, and regulatory policy. I then show how rapidly moving events in embryo research have built on those earlier settlements. Finally, I use my experiences on a U.S. ethics committee to describe, in ethnographic detail, how such committee work resolves puzzles at the borderlines of ontological and moral specification. I display the ways in which our committee's deliberations replayed familiar elements of U.S. political culture while addressing what to do with troublesome entities. In closing, I reiterate the value of restoring a comparative perspective to a terrain that has perhaps too quickly become the turf of a universalizing professional discourse (Evans 2006).

Distinctions that Matter

The influence of political culture on bioethics, and consequently on ontological politics, can be clearly seen in national characterizations, or recharacterizations, of the developing human embryo in response to new technologies. Scant decades after their invention, methods of generating and manipulating human embryos outside the female body are available worldwide. They include in vitro fertilization (IVF), cryopreservation, preimplantation genetic diagnosis (PGD), gene therapy, and human embryonic stem cell research. Services to produce, test, sort, and selectively implant IVF embryos can be purchased in fertility clinics from Manhattan to Mumbai and from Brussels to Beijing. Research methods and objectives have also diversified, raising the two issues flagged by President Bush in August 2001: Can we meaningfully distinguish among stages of embryonic development? Is there a moment at which a clump of cells—mere matter—becomes incipient life, part of our moral community, and hence entitled to the protections that law accords to personhood and human dignity?

These issues keep recurring as technologies evolve. In vitro embryos, the source of the original "test-tube babies," were almost naturalized through the rapid spread of assisted reproduction; as long as they are implanted into a mother's womb, they seem hardly different from embryos produced by procreation. But the techniques used to generate these extracorporeal precursors of human selves have been used to make less easily assimilated entities. The status of nonimplanted IVF embryos still remains deeply contested. Dolly was the result of somatic cell nuclear transfer (SCNT), a technique that removes the nucleus from a fully developed (somatic) cell taken from an adult body and implants it into a developing egg from another animal. The application of SCNT to clone human cells remains a flashpoint of concern. To defuse those anxieties, a few scientists have turned to another technique with uncertain ontological consequences: altered nuclear transfer, through which the somatic cell nucleus can be genetically altered to disable its full development (see Testa, chapter 4, this volume). Still another technique, potentially useful for preventing disease transmitted through the mother's mitochondrial DNA, places the nucleus from a zygote with mitochondrial defects into a healthy zygote whose nucleus has been removed. The resulting entity was quickly dubbed the "three-parent embryo" (Keim 2010).

Ethically, the problem in these cases is that science is not only discovering nature but remaking it (Rheinberger 1997). There is a famous passage in Plato's *Phaedrus* in which Socrates speaks of classification as a kind of expert deconstruction: a matter of cutting natural formations cleanly at the joints, just as an experienced butcher might carve up the body of an animal. Socrates was speaking of different kinds of love, but the passage has been interpreted as applying to natural kinds in general. In this vision, reality comes already divided into meaningful packets—long before we have the means to gain access to that reality. As Ian Hacking put it, "Quarks, the objects themselves, are not constructs, are not social, are not historical"; and again, "calling a quark a quark makes no difference to the quark" (Hacking 1999, 30, 105). The task of the ancient natural philosopher, as also of the modern scientist, is to recognize where nature placed the joints, and to perform the surgery that neatly displays nature's preordained classifications to fellow human beings.

But ontological surgery feels anything but foreordained in relation to the boundary-crossing products of biotechnology. Despite the best efforts of researchers and ethics commissions, by the turn of the twenty-first century only Britain among Western nations had officially carved embryonic life into two discrete segments for purposes of research and regulation. The first "natural" phase of development recognized in UK law consists of a period up to fourteen days, when the conceptus is neither regarded nor treated as human. In the second part, after fourteen days, the developing entity is seen, and treated, as protohuman life, with corresponding moral entitlements. Before fourteen days, many kinds of research on the embryo are fair game, under supervision of a legislatively created regulatory agency, the Human Fertilisation and Embryology Authority (HFEA).[1] Human embryos, the HFEA determined in 2004, can even be cloned and grown for research up to that cutoff point. After fourteen days, research is no longer permitted; the fourteen-day entity becomes from that point onward part of the human community.

The fourteen-day line was not biologically arbitrary. Developmental biologists had long recognized that two weeks is about the time when the so-called primitive streak appears, that is, the precursor to the nervous system. A biologist once told me that there is a chance, after this time, that the embryo may actually feel pain. Yet such speculations are not only scientifically untested, but also create no bright lines for moral action. It is not obvious, for instance, why the same reasoning that protects early embryos from research should not also apply to deter abortion after fourteen days, though that is not the subject of this chapter.[2] Nor is it self-evident how a biological hypothesis, such as the possible experience of embryonic pain, should link up with specific moral rights and obligations. Yet, under British law, a clear moral-metaphysical division was established at about two weeks into human development. What is mere matter before becomes in the eyes of law and policy a potential human after. How did this decision take hold, without either a strong scientific consensus or active public engagement?

Some explanations can be set aside, as I showed in earlier work (Jasanoff 2005a, 2005b). One is that political interests prevailed. British reproductive scientists, according to this account, gained a head start in embryo research because of their early successes with in vitro fertilization. The work of Robert Edwards, winner of the 2010 Nobel Prize in Medicine, and Patrick Steptoe had led in 1978 to the birth of Louise Brown, the world's first test-tube baby. Some have argued that British reproductive biologists, eager to maintain their comparative advantage, swayed Parliament into adopting a permissive regime for embryo research. Another explanation centers on the successful application of bioethics to a knotty problem. In 1984, the highly regarded Warnock Commission, chaired by Oxford philosopher Mary Warnock, advised the British government that embryonic development did not have to be seen as a continuous process: an ethically sustainable line could be drawn at fourteen days (Warnock 1985). On this account, ontological surgery succeeded here because the steady hand of experienced ethicists guided the knife blade.

But the lens of history is never kind to univocal explanations, and this case is no exception. Neither political interest nor bioethics satisfactorily accounts for British policy. Powerful scientific lobbies exist in many countries, including the United States, but they have not been able to translate their pro-research views into a firm national policy favoring embryo research. Various national ethics committees, as well, have tried but failed to produce consensus on the embryo's biological and moral status. Then, too, legislating the fourteen-day line in Britain was not straightforward. There was much uneasiness about this demarcation, and a victory for the government's position was not seen as assured. The Warnock report took an extraordinary six years to move from advice to law. In that time opposing pro-life groups managed to get their views accepted for a time by a majority in the House of Commons.[3] It remains puzzling, then, how a pragmatic solution, worked out in closed committee meetings, became a publicly ratified, *accepted* way of thinking about something so basic as when human life begins.

Authorizing Visions

To make sense of that puzzle, I have argued, we need to turn back the pages of British history to a key bioconstitutional moment in 1988 (Jasanoff 2005a, 152–155). It was during a debate in the House of Lords, when the government presented a White Paper containing its model embryology law, that British politics clearly displayed its particular approach to ontological surgery—its ability, in effect, to carve up nature and so to alter collective perceptions of the meaning of life. In the unlikely ambience of the unelected, unrepresentative House of Lords, Britain in effect solved one of life's deepest mysteries. Circumventing both the complexities of biology and the cacophony of politics, a line was drawn that became biologically, morally, and democratically self-evident.

We can identify in those proceedings a precise moment when categories were settled and ethical principles fixed. The crucial debate began with a healthy exchange about developmental biology among British peers: on one side, a sternly empiricist gaze that saw human development as continuously unfolding, with no clear points of demarcation; on the other, an insistence that experts can make distinctions that are not discernible though mere common sense. The two positions divided on whether the entity created through human fertilization is at first a preembryo, different in kind from the later, more complex, and ontologically distinguishable organism known as the embryo (Mulkay 1997).

Speaking for common sense, Lord Kennet, derided the very concept of a preembryo existing before fourteen days but not after: "Pre-embryo tells us nothing. A pre-embryo goes on to be an embryo, which is a pre-foetus, and a foetus, which is a pre-baby, goes on to be a baby: and a baby, of course, is a pre-adult. We do not call any other stage of human life story something beginning with 'pre.' It is a negative definition which merely says that it is not an embryo in order to avoid the stigma of destroying embryos."[4] Strangely, the Archbishop of York, representing religious authority, enrolled science to explain why he accepted the idea of discontinuous development: "There is a certain fluidity about the identity of the organism [the embryo] at this early stage, and that uncertainty is not resolved until the cells, instead of just going through the process of multiplying as happens in the very early stages, begin differentiating. It is only at that stage that one can begin to talk in any meaningful way about the embryo."[5]

Science and religion aligned on the same side against lay common sense? Clearly mediation was needed to find shared ground.

Fittingly, it was Mary Warnock herself who found the right form of words to bridge the gulf between empiricism and expertise. After shepherding history's most famous bioethics committee to a consensus, Warnock had risen rapidly in the British Establishment. At the time of the House of Lords debate, she was a baroness with a seat in the Lords, as well as Mistress of Girton College at the University of Cambridge. Warnock too pleaded for recognizing a distinction between the embryo and the preembryo, but her argument invoked a higher power than demotic sense or scientific authority: "There is a very noble prayer frequently uttered in my college chapel that we may be given the grace to distinguish things that differ. I, as a member of my college chapel, pray this most fervently because I believe that it is our moral duty to make a distinction in this case between the pre-embryo and the embryo which is to become the individual and to act on this distinction."[6] Very likely, as I have documented elsewhere, she was referring to a text attributed to the fifteenth-century Christian mystic Thomas à Kempis (Jasanoff 2005a, 154–155). The crux of that prayer is the insufficiency of human sight, and by extension of human knowledge, when it operates with no guidance but its own. In referring to Thomas à Kempis, Warnock performed, at one level, an act of profound self-abnegation. Unlike Lord Kennet and the Archbishop of York, she admitted that her own eyes might deceive her. Human sight, whether of the scientist, the ethicist, or the man in the street, she implied, may be biased, distorted, or incomplete unless it is guided by God's grace, which provides the only truthful account of things as they *are*.

In the Anglican Protestant world, God's truth—cutting nature at its divinely determined joints—can make itself known through secular philosophers and even ordinary people as well as through ordained clerics. But even God's grace needs secular authorization if it is to persuade a polity. In Britain, compelling accounts of the world as it is must be empirically grounded and attested to by appropriately authorized witnesses to achieve credibility.[7] For an ontological claim (this is how things *are*) not to appear idiosyncratic or subjective, let alone for people to accept it as morally binding (this is how things *should be*), the right sort of person has to make the claim in the right language and with the right kinds of evidence. The Warnock committee, for all its careful deliberation, had not been able to complete the translation from ethics to politics. That task fell again to Baroness Mary Warnock, who concentrated within herself the combined social capital of recognized public service (via the Warnock Commission), academic leadership (via Girton College and the University of Cambridge), and high political standing (via the House of Lords).

Other Visions, Other Settlements

Two other national cases, Germany and the United States, underscore not only the uniqueness of the British policy dispensation for embryos, but also the connections between ethical norms and political legitimation that account for cross-national divergences. Both of these countries also had to perform some sort of ontological surgery in order to place their research and funding policies on scientifically and politically acceptable footings. Neither, however, split embryonic development into distinct phases as did Britain. Germany followed the most restrictive course, extending human dignity to embryos from the moment of nuclear fusion between sperm and egg, and permitting no "spare" (not implanted) embryos to come into being through fertility treatment. The United States debated but did not resolve the status of the embryo at the national level. Instead, the market was permitted to govern reproductive technologies while the vagaries of national politics shaped federal support for hESC research.

The roots of these divergences can be traced to bioconstitutionalism: fundamental understandings about the state's responsibility for life that are articulated and enacted through law and policy. In Germany, there was no question that the state has an absolute duty to protect human dignity; the issue was how to map that obligation onto embryonic development. Here, opinions differed. The *Nationaler Ethikrat* (National Ethics Council, 2001–2008), for example, observed in relation to stem cells: "Although everyone agrees that the protection of human life is a primordial moral and constitutional concern, opinions differ on the protection to which human life is entitled during its early embryonic development."[8] That developmental period, in other words, is an ontologically and morally gray zone.

Under German notions of the rule of law, the state's prime responsibility in these unsettled circumstances is to maintain moral and legal certainty. This Germany accomplished largely by legislating ambiguous entities out of existence. The 1990 Embryo Protection Law, one of the most stringent of its kind in Europe, made it a crime to engage in activities that might open the door to the proliferation of preembryos. It provides that only as many IVF embryos should be created as can be implanted in one cycle into a woman undergoing treatment. Spare embryos do not (for the present) enjoy *any* valid ontological or constitutional status under German law. They are literally *non*entities, not entitled to physical or legal existence. Needless to say, the creation of embryos for research purposes is forbidden.

In the United States, the moral status of the embryo was debated loud and long in forums ranging from Congress to the Supreme Court, from presidential ethics commissions to the popular press, and in any number of books and articles by prominent scientists, lawyers, ethicists, and other public intellectuals. In America, too, the answer to the question when human life begins for policy purposes eluded definitive resolution. Both politically and legally, the status of the early embryo remains in doubt. It can be eliminated through abortion but not created through federally funded research.[9] It can be put up for "adoption" by willing gamete donors, registered for research in the state of California, and frozen for posthumous birth following a parent's death. As yet it cannot be bought or sold. The embryo's uncertain ontological status has not prevented human embryos from coming into being en masse. Created through IVF, but not implanted to induce pregnancy, hundreds of thousands of frozen embryos are held in storage facilities around the country, a population of legally indeterminate beings offering, as research material, untold promise of regeneration and rebirth.

What Britain addressed through empirical observation and politically authorized vision, and Germany through transcendent moral and legal principles (human dignity and the rule of law), the United States tackled through the market, that most flexible instrument of governance. In the marketplace of IVF, where most spare embryos originate, private clinics are free to determine many rules of the game governing the production, storage, use, and disposal of embryos, subject only to a patchwork of state laws and regulations. In the marketplace of scientific inquiry, different priorities and even different ethical rules apply to research with embryos and hESCs, depending on whether (and when) the work was funded, and by whom. And in the marketplace of policy, discrepant rule systems flourish side by side, according to whether the behavior to be regulated is under state or federal control. There has not been as yet a felt need for national uniformity—ontological, moral, or legal—as in the UK and German cases, even though clashes among competing value systems continually arise, as for example in the unexpected 2010 court decisions on the patentability of human genes and the legality of the Obama administration's stem cell guidelines.[10]

Ontological Surgery at Life's Frontiers

The responsibility structures laid down in the early years of debating the human embryo and, later, human embryonic stem cells, have exerted their hold on subsequent developments. Britain, Germany, and the United States continue to follow divergent paths in their official tolerance of research with embryos and their derivatives. The loci and discourses of ethical deliberation also differ, coproducing different descriptions of the same entities, as well as different judgments of responsibility toward them.

In Britain, the objectification of the pre-fourteen-day embryo became so taken for granted that, as Giuseppe Testa relates in chapter 4, the UK House of Lords unproblematically extended the HFEA's regulatory authority to embryos produced by other means. They were judged, in the end, similar things—or, in the eyes of the Law Lords, members of a single natural kind (see also Jasanoff 2005a, 200). By contrast, guidelines issued by the U.S. National Institutes of Health under President Obama—even if they survive legal challenge—do not allow federally funded research on embryos derived through somatic cell nuclear transfer or other techniques (NIH 2010, Section V). In Britain, bioethical advice has been tightly coupled to HFEA's work. The internally appointed Ethics and Law Committee (ELAC), consisting primarily of HFEA members, closely tracks the Authority's areas of concern (e.g., fertility treatment abroad, egg donation, and surrogacy). Though HFEA is frequently criticized for overregulation by embryo researchers and IVF advocates, its record of approving new modes of research is among the most liberal in the world, suggesting that the UK government is confident about its capacity to maintain moral order at the forefront of human biological research.

Although HFEA's decisions normalize borderline entities and practices as acceptable (or not)—as in its highly publicized September 2007 decision to allow research uses of "cybrid" embryos formed by inserting the nucleus of a human egg into an enucleated cow or rabbit egg (HFEA 2007)—much of the public work of ontological surgery happens in the talk of researchers. Strikingly, these scientists appear to have thoroughly internalized the "object" status of the early embryo, translating Baroness Warnock's revealed distinction into their own commonsensical terms. Thus, when HFEA announced in June 2004 that it was about to approve the cloning of human embryos for research, Alison Murdoch, one of Britain's most prominent fertility scientists and later head of the Newcastle Fertility Centre, stepped in briskly to defuse concern. In an interview with the *Sunday Observer*, she said: "Out of ten eggs produced during IVF treatments on average seven are used. The other three are spare and would otherwise have been disregarded. We are not trying to clone a baby. . . . These embryos have no more moral status than blood taken from a patient" (Barnett and McKie 2004, 1). No troubling moral lines would be crossed, Murdoch implied, for the simple reason that the researchable embryo was not the kind of thing we need to worry about. It is a form of biological waste—nothing more.

In April 2010, Douglas Turnbull of Newcastle University used even more explicitly matter-of-fact language to describe the production of the so-called three-parent embryo (see previous mention). "What we have done is like changing the battery on a laptop," he said, adding, "The energy supply now works properly but none of the information on the hard drive has been changed" (Boseley 2010). Turnbull's choice of metaphor reflects a seamless and total translation of life into information without the speaker seeming to register any possibly jarring ethical resonances.

Such public talk is, if course, impossible in Germany, because the legal environment for research does not permit the production of many novelties that have emerged from British and American clinics and laboratories. The spare IVF embryos that constitute the raw material for this mode of production have no right to exist in the German bioconstitutional regime. The German state, as we have seen, ensured that these ontologically and ethically uncertain things, or "monsters" (Jasanoff 2005a, 158; 2005b), would not come into being within German territory. Against this backdrop, a July 2010 decision of the Federal Court of Justice (*Bundesgerichtshof*, or *BGH*) holds special interest because it opened the door a crack in the direction of greater intervention into embryonic life (BGH Judgment of July 6, 2010, 5 StR 386/09). The logic of the decision, however, left undisturbed the legislated consensus that embryos should be accorded human dignity.[11]

The case involved a Berlin fertility doctor who had conducted preimplantation genetic diagnosis (PGD) in three cases in order to determine which IVF embryos possessed genetic defects that might prevent the birth of a healthy baby. Potentially, such testing and subsequent decisions not to implant the defective embryos exposed the doctor to criminal penalties under the 1990 law.[12] In concluding that his actions were not legally forbidden, the BGH looked, as courts must, to the German Parliament's intent. Noting that PGD was neither foreseen nor provided for at the time of the law's passage, the BGH concluded that the technique did not violate any of the German Parliament's cardinal prohibitions: instrumentalizing embryos for purposes other than inducing a pregnancy, exposing embryos to needless risk of destruction, promoting surrogacy, sex selection for reasons other than fetal health, or selection for other desirable genetic traits.

The starting point of BGH's analysis was that, under German law, all embryos at whatever stage of development belong to the human moral community. Accepting this continuity as given, biological arguments became significant for the court only to establish that PGD testing at the 40–80 cell stage poses no undue risk to the blastocyst (para. 23). Far from causing harm, the court reasoned, PGD serves the legally sanctioned and socially desirable goals of preventing the birth of a child suffering from severe birth defects (para. 25) and protecting the pregnant woman against undue burdens resulting from such a birth (para. 26). In sum, If PGD is allowed, it is precisely because it appears to be animated by the same respect for embryos and their social relations that the law already recognizes in connection with childbearing—where it is the state's duty to encourage successful pregnancies and prevent the birth of doomed or severely handicapped babies.[13]

If law and regulation in Britain tracked the Warnock Commission's initial ontological judgments, and if ethics in Germany followed the parliament's biopolitical intuitions, what of the United States, where it has proved almost impossible to find common ground concerning the *is* and the *ought* of early human life? Perhaps predictably, in a laissez faire culture there is no single answer to this question. Rather, as Hurlbut (2010) relates in his account of Proposition 71 (California's controversial 2004 stem cell ballot initiative), there is an ongoing struggle over who has the right to attach controlling meanings to entities such as embryos, stem cells, and other derivatives. Scientists, nongovernmental groups, and even private companies may get into that act, as the following example illustrates.

In November 2001, Advanced Cell Technology (ACT), a Massachusetts-based company specializing in regenerative medicine, announced that its scientists had successfully cloned human embryos. Like Britain's Alison Murdoch, Michael West, then the company's CEO and Chief Scientific Officer, engaged in ontological surgery to defuse moral and political criticism. "We're talking about making human cellular life," he said, "not a human life. A human life, we know scientifically, begins upwards, even into two weeks, of human development, where this little ball of cells decides, 'I'm going to become one person or I am going to be two persons.' It hasn't decided yet."[14] Somewhere around fourteen days, West intimated, the power of reasoning enters into the developing embryo. Before that time, the indecisive "little ball of cells" is only matter; afterward, it is on the way to acquiring personhood. Key here was West's assertion, consistent with well-known patterns of American civic epistemology, that "we know scientifically" when human life begins. When science speaks clearly, even the CEO of a for-profit concern can appropriate its authority to underwrite potentially controversial distinctions and classifications.

West was a controversial figure, and by 2007 even ACT was at pains to clothe its talk of breakthroughs in the language of moral compliance. The company announced the creation of the first hESC line "without destroying an embryo," thereby addressing the ethical concerns raised by President Bush's 2006 veto message (ACT 2007). Increasingly, however, the task of normalization falls in the United States to ethics committees, whether attached to private companies or to universities. We turn now to an example of the work done by such bodies.

Vignettes from the Field: Ontological Puzzles in an ESCRO Committee

Continued legal and political uncertainty in the United States means that much of the task of ontological and moral clarification falls in practice to ethics committees. Such bodies can be highly visible, as in the case of presidential ethics commissions, or virtually unseen, as in the case of Institutional Review Boards concerned with human subjects research and, more recently, Embryonic Stem Cell Research Oversight (ESCRO) committees. Of these, it is the latter that have the most direct implications for research; these relatively low-level bodies enjoy considerable authority to accept, modify, or in exceptional cases block research protocols. How institutional ethics committees form their judgments is accordingly a matter of considerable significance for democratic politics. In this section, I offer an analysis of one such committee based on my own observations as a participant. Despite their lack of overt democratic accountability, I suggest, U.S. ethics committees achieve authoritative ontological settlements partly by adopting tacit, culturally accepted norms of pluralist deliberation and associated notions of proof and persuasion (Jasanoff 2005a, 247–271).

In 2006, I was appointed to serve along with a dozen or so colleagues on Harvard University's ESCRO Committee. These committees, which exist wherever federal funds are expended for stem cell research, are not required by law but were formed in response to public unease over hESC research. Seeing a threat to science, the National Academy of Sciences— the most prestigious scientific body in the United States—issued a report on the scientific background and ethical implications of this research domain, together with guidelines to ensure that research using hESCs would be carried out in a responsible manner. Stem cell scientists, the Academy noted, should "be sensitive to public concerns about research that involves human embryos" (NAS 2005, 100). To this end, the guidelines asked each institution involved in hESC research to form an ESCRO committee. Any English speaker will recognize the homonymy of the abbreviation ESCRO with the legal term "escrow"—referring to money safeguarded by a third party for a specified purpose until some stated condition is fulfilled. The very name of these oversight committees, then, nicely merges the ideas of commodity and of holding in trust. Biopolitics meets and melds with the bioeconomy and its promise of public benefits from research.

A committee that operates outside the requirements of the law is a particularly interesting site for observing the influence of bioconstitutionalism: it is a private sphere, in that it is not governed by the state, and yet it is concerned with the legitimacy and enforceability of its decisions. I offer a few examples to illustrate the committee's dedicated efforts to discharge its duties, how it understands its representational role, and how it deals with ambiguity and ambivalence in governing life.[15]

According to the NAS guidelines, and conforming to widely accepted criteria for constructing expert advisory bodies in the United States (Jasanoff 1990), ESCRO committees are conceived as representative of relevant scientific and technical expertise as well as diverse public values held about such work. The guidelines state, "The committee should include representatives of the public and persons with expertise in developmental biology, stem cell research, molecular biology, assisted reproduction, and ethical and legal issues in hES cell research. It must have suitable scientific, medical, and ethical expertise to conduct its own review" (NAS 2005, 100). The named standpoints are imagined as covering all significant positions with respect to the merits of hESC research: science(s), ethics, law, and the public. Consistent with the dynamics of pluralism, a working consensus is supposed to develop through exchanges among informed stakeholders.[16] In practice, the numbers distinctly favor technical expertise, producing an implicit hierarchy between facts and values. In effect, the first and most important tacit ethical norm applied by ESCRO committees is that the work should be adjudged "good science." As long as it passes that test, the presumption (often stated by members) is that our job is to make ESCRO review as efficient as possible, consistent with ensuring that the research complies with applicable guidelines.

As a representative body, the ESCRO committee interestingly problematizes the very meaning of representation, because the relationship between members and who or what they speak for is inexplicit and, to some degree, subjective. Some members seem to believe they are there primarily to defend the interests of hESC researchers, including their own colleagues at the university. Their interventions routinely seek to exclude restrictions that they believe would not make sense to the (occasionally named) researchers they feel they are representing. Others believe that they are speaking for the integrity of their field of expertise. One member, for example, is convinced that there is nothing ethically specific enough about stem cells to justify constraints on hESC research; this member routinely objects when committee recommendations appear to reinforce the apparent specialness of these entities. Still others believe that they are there to represent the interests of the donors who gave their spare embryos for research. Much of their input takes the form of trying to imagine what donors would say if they were approached for additional information or expanded permissions. In sum, the ESCRO committee seems constantly to be imagining and speaking for groups within a shadow polity—of scientists, ethicists, and embryo donors—toward whom they feel personally responsible. Committee discourse is that of answering to these virtual constituencies rather than to abstract moral principles.

Discussions of this sort can quickly get mired in minute technical details as members try to imagine the everyday transactions that embryos, donors, and research groups might get mixed up in. Worries about maintaining donor anonymity are especially common. I once spent an hour in an ESCRO committee meeting trying to decide whether an absent donor's consent form needed to be notarized (i.e., certified by an official notary), which would reveal the donor's identity to the notary, or whether it would be sufficient to require copies of relevant documents to be faxed to the research group without a notarized signature. Another time, we deliberated whether canisters containing frozen embryos preserved in liquid nitrogen could be opened up to replenish the gradually evaporating nitrogen, even though there was a small risk that the label bearing the identity of the donor might be revealed during the process. Someone suggested that enough gaseous vapor would swirl around the canister opening to make the small identifying labels on the rims unreadable. Such moments are easy to ironicize, but they represent deeply serious, imaginative exercises in representation, putting the committee in the shoes of potentially real actors confronted with the real and messy details of technoscientific practice.

Sometimes the committee does address grander issues. In 2007 we hosted a small, invitation-only workshop to discuss the ethics of experimentation involving human-animal chimeras. These entities are produced when pluripotent human cells, including stem cells derived from embryos, are inserted into other animal species. One experimental protocol involves introducing human neuronal (nerve) cells into mouse brains, creating a distant but troubling risk that treated "human neuronal mice" might develop aspects of human consciousness. The NAS (2005, 105) guidelines for hESC research explicitly ask that protocols combining stem cells with nonhuman embryos and fetuses should be submitted for ESCRO review. Our public meeting, prompted by a then-pending research protocol at Harvard, aimed to explore the sorts of ethical issues such experiments might raise. For a brief few hours on a June afternoon, we held a conversation in which liberal and communitarian arguments on the subject were aired, but the dynamics of the meeting subtly ensured that the liberals had the last word.

Importantly, everyone at the workshop approached the meeting with an open mind, acknowledging that there were uncertainties, both ontological and moral. It was clear from the way people talked about possible scenarios that these studies caused considerable anxiety. For example, although most agreed that the correct protocol was to destroy immediately any chimeras that begin displaying "nonmouse behavior," no one could answer why we should kill an organism that was beginning to resemble humans more closely. People joked awkwardly to cover their unease. One said animal care workers in the lab should be carefully trained so they would know what to do if they came in one morning and found a lab mouse smiling or reading the newspaper. Another, an ethicist, spoke against sacrificing such highly developed organisms and recommended laughingly that they be sent instead on a "Club Med" vacation. Others spoke of running tests of self-consciousness—such as psychology's well-known "mirror test" to see whether an animal recognizes itself in a mirror (Gallup 1970)—to judge whether a line of impermissibility was being crossed. But speakers admitted on questioning that they had no idea what kinds of mouse behavior would give evidence of breaching an ethically meaningful line.

Repeatedly, participants referred to chimerical mouse characters in two well-known American children's books: E. B. White's 1945 classic *Stuart Little* and Robert C. O'Brien's (1971) Newbery Medal–winning work, *Mrs. Frisby and the Rats of NIMH*. The eponymous hero of *Stuart Little* is a mouse born into a family of human New Yorkers, and is fundamentally a human in a mouse body. In *Mrs. Frisby*, by contrast, the rats of the title are unquestionably rats, but they have acquired superhuman intelligence, as well as reading, writing, social structure, and technological skills, as subjects of lab studies at the National Institute of Mental Health. Fiction served ethics by disentangling two scenarios: first, that research mice become human in all but bodily form; second, that mice remain mice, but begin to display aspects of human consciousness. Workshop participants seemed to agree that their ethical judgments would depend on which imaginary the chimeras would conform to, but which was the more troublesome prospect, and why ethical thought should distinguish between the two possibilities, was never clearly articulated.

Two speakers attempted to shift the focus of debate from the risks of accidental "humanization" to a broader, more communitarian consideration of the values at stake in hybridizing humans and animals. Joseph Vining, professor of law at Michigan, eloquently warned against the "eugenic temptation" that foregrounds certain human endowments as favored (in this case, consciousness and cognitive capacity) and discriminates against others on that basis (see Vining 2008).[17] Donna Haraway, historian and philosopher of biology at the University of California at Santa Cruz, spoke to the meeting by telephone about the impossibility of finding any valid biological distinction between humans and other animal species. Like Vining, Haraway stressed the importance of thinking ethically about our relationship with animals and situating policies against a backdrop of such analysis.

In the end, however, a more utilitarian mood prevailed, helped along by the order of presentations. The chair grouped three professional ethicists to offer their comments together at the meeting's end. All three agreed that chimera experiments should continue in spite of the uncertainties because, as one speaker put it, it is the university's duty to maintain a free space for research, and not to take sides between scientists' (and presumably the public's) divergent values. In the resulting liberal environment, he argued, those who want to do research at the frontiers of neurobiology would be free to use human-animal experimental systems; those who do not want to conduct such research can opt out. This reasoning essentially eviscerated the notion of collective ethical responsibility for framing research agendas, let alone toward humanmade hybrids, because it relegated ethics itself to the realm of private belief and individual choice. Scientific inquiry, by contrast, was presented as an unquestionable public good, to be promoted in the "free" intellectual space of a university. Closing on this note, the meeting appeared to endorse research using human-animal chimeras more through a kind of collective body language than through collective reason.

My own role on this committee deserves a few words of reflection, because, by disciplinary training in science and technology studies, I am necessarily observer as well as participant, critic as well as policy actor, and expert as well as lay recipient of information from knowledgeable others, and thus a player on multiple levels at once. How should a science studies expert with training in reflexive analysis of facts and claims, as well as with academic interests in democratic theory, responsibly contribute to a body like this? In part, when issues of committee role and representation arise, I try to supplement well-known and well-represented interest group perspectives with a more holistic sense of our responsibility as a deliberative body. I have tried on occasion to speak for an emergent role in which we might seek, through our conversations, to unravel the issues that a broader public might find at stake in hESC research if they knew everything that we learn through our serious and respectful attempts to educate one another. In short, I try to inject forms of advice-giving that I describe as culturally "German" in my comparative research (Jasanoff 2005a)—aware of the obstacles that such attempts are likely to encounter in an environment where they run against the grain of pluralist deliberative practice.

The committee is generally respectful of these efforts, provided they are coupled to pragmatic recommendations. For example, our minutes became considerably more detailed when I suggested that future researchers might have an interest in studying our evolution. Sometimes, alliances build within the committee among those who favor greater caution or more accountability, though on grounds ranging from stakeholder interests in privacy or gender to my own more abstract interest in deliberative norms. In keeping with this subgroup's wishes, leading hESC scientists have occasionally appeared before the committee to explain the purposes and progress of their experiments, despite one or two objections from committee members that this might impose undue burdens on research. Of course, pragmatic considerations usually prevail. One time, when I tried to initiate a discussion on the ESCRO committee's responsibilities as a surrogate democratic forum, an influential committee member responded, with genuine warmth, "I love it." But, he went on to say, there was very little time left, and we had some research protocol approvals still pending.

Conclusion

Recent advances in the life sciences and technologies have created a testing ground for bioconstitutionalism. No longer limited to discovery, the biosciences are a source of novel constructs and practices that challenge settled understandings of what the German National Ethics Council called the state's "primordial moral and constitutional concern" for human life. Against this backdrop, I have argued, bioethics plays a metaphysical as well as a moral role. By sorting new entities (and sometimes old ones) into ethically manageable categories, bioethics helps define the ontologies, or facts of life, that underpin legal rights and condition scientific and social behavior. In short, bioethics as much as science itself performs ontological surgery; it is only when science and ethics are at work together that the "joints" of biotechnologically manipulated nature stand publicly revealed.

The puzzles confronted by ethical decision makers in Britain, Germany, and the United States illustrate the uncertainties and confusions that accompany all attempts to make, or remake, the facts of life. Which new things (IVF or SCNT embryos, "cybrids," or human neuronal mice, for example) should we admit into our actual and imagined moral communities, and under what structures of responsibility and governance? Deliberation on such issues, I have argued, is situated within national bioconstitutional dispensations that crucially shape both reasoning and outcomes. Bodies such as Britain's HFEA, Germany's *Bundesgerichtshof*, and U.S. ESCRO committees are places where, under the rubric of making ethical judgments, political theories are tacitly argued, culturally dominant views about representation and public reason shape debates, and ethical principles follow from basic choices about which positions will be represented and by whom. Those deliberations in turn shape our conceptions of entities at the frontiers of life: what a thing is and how we should treat it are repeatedly, if diversely, resolved together.

In Britain, an early elite consensus on the "preembryo" as empirically not human set the baseline for subsequent ethical, legal, scientific, and public discourse. The result has been a tightly regulated but relatively permissive research environment, in which the fourteen-day bright line has held fast against expressions of moral concern. In Germany, overarching commitments to human dignity and legal certainty prevented boundary-crossing entities from officially coming into being. The PGD case oddly illustrates that experimentation with totipotent embryonic cells may be legally more problematic than testing and discarding blastocysts with genetic defects, which the high court swept into a medicalized discourse of caring for mothers and newborns. In the U.S. ESCRO committee, by contrast, ethical questions are framed largely in terms of risk, consistent with a national commitment to liberalism in science as well as commerce. That framing in turn privileges free scientific inquiry and individual access to reproductive services over relational questions of collective responsibility for new biological objects. The ethics committee's own legitimacy rests on the bedrock of pluralist politics, which takes ontologies and interests as given, and leaves little space for debating transcendental questions of *is* or *ought*.

Modern science has laid out no singular path for classifying or ordering the constructs of biotechnology. Science does not tell what we shall do and how we shall arrange our lives any more than it did when Max Weber ([1918] 1948) quoted Leo Tolstoy in "science as a vocation." Bioethics, too, emerges as profoundly context-dependent. The three national approaches to dealing with early embryonic development described above testify to persistent cross-cultural differences in bioconstitutionalism even among closely related, rational, democratic societies. These divergences deserve attention precisely because each represents an ongoing grand experiment in taking responsibility for life. Comparison allows us to recover, through contrasts among several active lifeworlds, issues that may have been sidelined or silenced without explicit consideration in each. It destabilizes our own familiar, taken-for-granted categories, and thus invites us to be more self-aware about the foundations of our bioethics.

Notes

1. It is illustrative of Britain's culture of elite formation that Lisa Jardine, the distinguished Renaissance historian of science and professor at University College London, was appointed HFEA chair from 2008 to 2011. Along with Jardine's impressive credentials as scholar and public intellectual comes lineage: she is the daughter of the famous scientist and science publicist Jacob Bronowski. In an interview shortly after her appointment, Jardine made it clear that "it isn't her role to have views. It's to make 'intellectual judgments' about complex issues on the basis of listening to people" (*The Times* 2008). By contrast, in 2009 Barack Obama appointed Amy Gutmann, professor and president of the University of Pennsylvania, to chair his Presidential Commission for the Study of Bioethical Issues. Gutmann is a leading philosopher and democracy theorist, presumably selected in part for her expertise in normative analysis.

2. In Germany, where embryos are accorded human dignity from the moment of fertilization, the argument for abortion rests importantly on the woman's countervailing interests. Full discussion of this issue would take us beyond the purposes of this chapter.

3. Warnock herself alluded to this difficulty nearly two decades after the passage of the law: "The research on which [IVF] depended was difficult and, controversially, involved the destruction of human embryos. So it was hardly surprising that the 1990 Human Fertilisation and Embryology Act, which covered the research and clinical practice, reached the statute book by the skin of the teeth of those, such as myself, who pushed it through" (Warnock 2008).

4. Hansard House of Lords Debate, January 15, 1988, col. 1497.

5. Hansard House of Lords Debate, January 15, 1988, col. 1461–1462.

6. Hansard House of Lords debate, January 15, 1988, col. 1470.

7. It has long been noted that scientific claims must be authoritatively witnessed if they are to gain credibility, and that much of the work of science is directed toward creating the preconditions of credibility (Shapin and Schaffer 1985). Manufacturing political credibility requires additional and different kinds of work, even when political authorities support their positions on the basis of scientific claims. See Jasanoff 2005a, chap. 10, on British civic epistemology.

8. Nationaler Ethikrat, Stem Cell Research, <http://www.ethikrat.org/_english /main_topics/stem_cell_research.html> (accessed August 2010).

9. The Dickey-Wicker Amendment, annually attached to appropriations bills for Health and Human Services from 1997 onward, prohibits the expenditure of federal funds for any research in which a human embryo is created, damaged, destroyed, or subjected to risk of injury or death beyond the extent already permitted by federal law. The ban covers the derivation of stem cells from embryos.

10. *Association for Molecular Pathology et al. v. United States Patent and Trademark Office*, 702 F. Supp. 2d 181 (S.D.N.Y. 2010), invalidated patents on genes related to breast and ovarian cancer held by Myriad Genetics on *Sherley v. Sebelius*, 2010 U.S. Dist. LEXIS 86441 (D.D.C. August 23, 2010), ruled that the NIH stem cell guidelines were contrary to applicable law.

11. For an English language account, see Tabke 2010.

12. The law mandates jail sentences of up to three years for anyone who uses an embryo "generated outside the body" for "a purpose not aimed at its preservation."

13. German conservatives immediately read the decision as pointing to an unacceptable gap in the coverage of the Embryo Protection Act and vowed to overturn the court ruling through legislative amendment. At the 2010 party conference of the Christian Democratic Union (CDU), Chancellor Angela Merkel called for a reinstatement of the ban on PGD.

14. "West: 'I'm just trying to help people who are sick,'" November 25, 2001, CNN .com, <http://archives.cnn.com/2001/TECH/science/11/25/cloning.west.cnna/>.

15. My account of discussions within the committee follow the ground rules laid down early in ESCRO's deliberative process that nonconfidential matters may be publicly disclosed, provided that they are not attached to named speakers.

16. Interestingly, this is the perhaps only committee I have served on at Harvard or elsewhere in which women significantly outnumber men. An alien visitor might conclude that the ethics of science is largely women's work!

17. In an essay published after the Harvard workshop but covering the material he presented there, Vining ironically dissected a report by a National Research Council working group that concluded there were no ethical objections to hybridizing human-nonhuman primates based on unnaturalness or crossing species boundaries. Vining questioned this group's right to speak as a representative body: "The pronouns 'we' and 'us' appear frequently in this Report. But from what stance are these beings speaking? How confident they are that they themselves are human?" (Vining 2008, 64–65). Questions about whom *we* are representing have never come up for explicit debate during our ESCRO committee deliberations, although tacit norms of balance prevail, as in the routine practice of having each protocol reviewed by a technical and a nontechnical member. Those norms are consistent with the kind of pluralist balance envisaged by the Federal Advisory Committee Act and in the practices of U.S. expert committees more generally (Brown 2009; Jasanoff 1990).

References

Advanced Cell Technology (ACT). 2007. "Advanced Cell Technology Develops First Human Embryonic Stem Cell Line without Destroying an Embryo." Press release. Retrieved from <http://www.advancedcell.com/news-and-media/press-releases/advanced-cell-technology-develops-first-human-embryonic-stem-cell-line-without-destroying-an-embryo/> (accessed December 2010).

Barnett, Antony, and Robin McKie. 2004. "UK to Clone Human Cells." *The Observer*, June 13, p. 1.

Boseley, Sarah. 2010. "Scientists Reveal Gene-Swapping Technique to Thwart Inherited Diseases." *Guardian*, April 14.

Brown, Mark. 2009. *Science in Democracy: Expertise, Institutions, and Representation*. Cambridge, Mass.: MIT Press.

Bush, George W. 2001. "President Discusses Stem Cell Research." Retrieved from <http://georgewbush-whitehouse.archives.gov/news/releases/2001/08/20010809-2.html>.

Bush, George W. 2006. Veto Message to the House of Representatives, July 19. Retrieved from <http://www.nytimes.com/2006/07/19/washington/text-stem.html> (accessed August 2010).

Callon, Michel, Pierre Lascoumes, and Yannick Barthe. 2009. *Acting in an Uncertain World: An Essay on Technical Democracy*. Cambridge, Mass.: MIT Press.

Evans, John. 2006. Between Technocracy and Democratic Legitimation: A Proposed Compromise Position for Common Morality Public Bioethics. *Journal of Medicine and Philosophy* 31 (3): 213–234.

Foucault, Michel. 1998. *The Will to Knowledge*. Vol. 1: The History of Sexuality. London: Penguin.

Gallup, Gordon G., Jr. 1970. Chimpanzees: Self-Recognition. *Science* 167:86–87.

Gottweis, Herbert. 1998. *Governing Molecules: The Discursive Politics of Genetic Engineering in Europe and the United States*. Cambridge, Mass.: MIT Press.

Hacking, Ian. 1999. *The Social Construction of What?* Cambridge, Mass.: Harvard University Press.

Human Fertilisation and Embryology Authority. 2007. "HFEA Statement on Its Decision Regarding Hybrid Embryos." Retrieved from <http://www.hfea.gov.uk/455.html> (accessed August 2010).

Hurlbut, James Benjamin. 2010. *Experiments in Democracy: the Science, Politics and Ethics of Human Embryo Research in the United States, 1978–2007*. Dissertation in the History of Science. Cambridge, Mass.: Harvard University.

Irwin, Alan. 2001. Constructing the Scientific Citizen: Science and Democracy in the Biosciences. *Public Understanding of Science* 10 (1): 1–18.

Irwin, Alan, and Mike Michael. 2003. *Science, Social Theory and Public Knowledge*. Buckingham, UK: Open University Press.

Jasanoff, Sheila. 1990. *The Fifth Branch: Science Advisers as Policymakers.* Cambridge, Mass.: Harvard University Press.

Jasanoff, Sheila. 2005a. *Designs on Nature: Science and Democracy in Europe and the United States.* Princeton: Princeton University Press.

Jasanoff, Sheila. 2005b. In the Democracies of DNA: Ontological Uncertainty and Political Order in Three States. *New Genetics & Society* 24 (2): 139–155.

Jasanoff, Sheila, and Sang-Hyun Kim. 2009. Containing the Atom: Sociotechnical Imaginaries and Nuclear Regulation in the U.S. and South Korea. *Minerva* 47 (2): 119–146.

Keim, Brandon. 2010. "3-Parent Embryos Could Prevent Disease, But Raise Ethical Issues." *Wired Science*, April 14. Retrieved from <http://www.wired.com/wiredscience/2010/04/mitochondria-engineering/> (accessed August 2010).

Mol, Annemarie. 1999. Ontological Politics: A Word and Some Questions. In *Actor Network Theory and After*, ed. John Law and John Hassard, 74–89. Oxford: Blackwell.

Mulkay, Michael. 1997. *The Embryo Research Debate: Science and the Politics of Reproduction.* Cambridge: Cambridge University Press.

National Academy of Sciences. 2005. *Guidelines for Human Embryonic Stem Cell Research.* Washington, D.C.: National Academies Press.

National Institutes of Health. 2010. Guidelines on Stem Cell Research. Retrieved from <http://stemcells.nih.gov/policy/2009guidelines.htm> (accessed August 2010).

O'Brien, Robert C. 1971. *Mrs. Frisby and the Rats of NIMH.* New York: Atheneum Books.

Post, Robert, and Reva Siegel. 2007. *Roe* Rage: Democratic Constitutionalism and Backlash. *Harvard Civil Rights-Civil Liberties Law Review* 42:373–433.

Rheinberger, Hans-Jörg. 1997. *Toward a History of Epistemic Things: Synthesizing Proteins in the Test Tube.* Palo Alto, Calif.: Stanford University Press.

Shapin, Steven, and Simon Schaffer. 1985. *Leviathan and the Air-Pump: Hobbes, Boyle, and the Experimental Life.* Princeton: Princeton University Press.

Skloot, Rebecca. 2010. *The Immortal Life of Henrietta Lacks.* New York: Crown.

Tabke, Erika. 2010. "German High Court Okays Preimplantation Genetic Diagnosis for IVF Embryos." dotIVF, July 7. Retrieved from <http://www.dotivf.com/forums/content.php?547-German-High-Court-Okays-Preimplantation-Genetic-Diagnosis-for-IVF-Embryos> (accessed August 2010).

The Times. 2008. "Lisa Jardine on Surviving Cancer and Leading the HFEA." February 9.

Vining, Joseph. 2008. Human Identity: The Question Presented by Human-Animal Hybridization. *Stanford Journal of Animal Law and Policy* 1:50–68.

Warnock, Mary, ed. 1985. *A Question of Life: The Warnock Report on Human Fertilisation and Embryology.* Oxford: Basil Blackwell.

Warnock, Mary. 2008. "Parliament Must Retain Moral Authority over Science." *The Observer*, January 13, p. 35.

Weber, Max. [1918] 1946. Science as a Vocation. In *From Max Weber: Essays in Sociology*, trans. and ed. H. H. Gerth and C. Wright Mills, 129–156. New York: Oxford University Press.

White, Elwyn B. 1945. *Stuart Little*. New York: Harper Trophy.

Wilmut, Ian, A. E. Schnieke, J. McWhir, A. J. Kind, and K. H. S. Campbell. 1997. Viable Offspring Derived from Fetal and Adult Mammalian Cells. *Nature* 385:810–813.

4

More than Just a Nucleus: Cloning and the Alignment of Scientific and Political Rationalities

Giuseppe Testa

"It's so scary to me that this guy I don't even know can do that. It's like he's killing me" (Maienschein 2003, 6). Tessa Wick, an American girl affected with diabetes, is talking about Senator Brownback's proposal to outlaw reproductive and therapeutic cloning. She is arguing that she can use one of her skin cells and make it into "something" (a cloned embryo, for example) from which to derive cells that could one day cure her disease. Somatic cell nuclear transfer (SCNT), the technology that generated Dolly (Wilmut 1997), entails the transfer of the nucleus from a somatic cell into an enucleated egg, which can then develop into an organism whose cell lineages are virtually genetically identical, and thus immunologically compatible, with the original nucleus donor. Hence, Tessa Wick is asking the Senate committee to reframe cloning as a right growing from her control over her own body. Just as she is entitled to use a piece of skin from her leg to repair a wound on her hand, she wants to be able to reprogram one of her cells and make it capable of producing insulin. The fact that this "thing" could, under certain circumstances, acquire a life of its own, intersecting with the trajectory of nascent human life, is irrelevant for her. This thing is hers by right: it comes from her and goes back into her.

In this chapter, I look at three examples of how cloned cells were enabled as socially legitimate scientific objects: the decision of the British House of Lords (HL) in 2003, the proceedings of the Italian Dulbecco Commission in 2000, and the proposal developed by the U.S. President's Council on Bioethics (PCB) to overcome the ban on public funding for human cloning. My aim is to gain insight into the practices through which science and society engage in the simultaneous production of knowledge and social norms (including constitutional rights), and how they go about ordering new living things within existing or yet to be invented categories.

Tessa Wick imagines around, and indeed inside, herself a space of freedom and individual autonomy over the reprogrammable states of her genome (clones, embryonic stem cells, and the like). We take her stark words as a starting point because they capture at once the full extent of this imagined space and the saliency of the controversy that pervades it. The task for the analyst is then to understand how different political systems and scientific cultures conceive and map that space, through which tools, and on what assumptions. The three cases I compare show that this mapping process is much more than a simple regulatory effort. Decisions about where to draw the boundary between individual freedom and collective responsibility, and about if and when clones exit Tessa Wick's imagined space to acquire public standing, shape and are shaped by ontological questions about the biology of the clones and epistemological assumptions about the technology that brings them into being.

Clones are prototypical of what Rabinow proposed as the hallmark of modern, science-based rationality: "The object to be known—the human genome—will be known in such a way that it can be changed" (Rabinow 1996, 92). In the case of cloning, we may well say that the object is known only because it can be changed. This makes the artifact an even stronger example of the kind of conflation through which "representing and intervening, knowledge and power, understanding and reform, are built in, from the start, as simultaneous goals and means" (92). If the philosopher-emperor Marcus Aurelius prompted us to contemplate the product of SCNT and ask 'What is it in itself? In its own constitution?' (Marcus Aurelius 2005, VIII), it would become clear that the SCNT product does not have its own constitution. Is it important where it is coming from? Or is it important where it is going, its potential to intersect with and reproduce life?

The three cases I examine took place in different institutions and discourses, but each has specific significance in the cloning debate and each points to different modalities of coproduction (Jasanoff 2004, 1–45). The relevance of the British case lies in the fact that it was the first example in which the science and technology of cloning passed through all levels of a judicial system. The report on cloning by the Italian DC—to my knowledge the first political body to frame clones as distinct from embryos—deployed the same argument as in the British example but reached opposite conclusions. The third concerns the physical coproduction of a novel cloning procedure and a novel laboratory entity through the joint work of the PCB and the MIT labs of molecular biology.

The United Kingdom: Reconstructing Embryos by Law

The problem of defining the entities derived by SCNT rose to public salience in the United Kingdom in 2003. A report by the Chief Medical Officer's Expert Group, published in June 2000 (Donaldson Report 2000) had considered the medical potential of SCNT and recommended that research using SCNT embryos should be permitted within the regulatory framework of the Human Fertilisation and Embryology Act of 1990 (HFEA). The government accepted these recommendations in August 2000 and promised to implement the recommendations. In November 2000, Bruno Quintavalle, on behalf of the Pro-Life Alliance (PLA), an antiabortion organization, petitioned for a judicial declaration that the SCNT embryos did not fall within the definition of embryo of the HFEA (*Quintavalle v. Secretary of State for Health* [2001]).

The High Court accepted the PLA's claim and declared that "human embryos created by cell nuclear replacement technique that involves the insertion of the nucleus of a human somatic cell into an unfertilized egg (oocyte) which had had its own nucleus removed are not 'embryos' within the meaning of s 1 of the Human Fertilisation and Embryology Act 1990 and, therefore, are not subject to regulation under the Act." The crucial Section 1(1) of the HFEA, upon which the decision was based, stated that (a) "embryo" means a live human embryo where fertilization is complete, and (b) references to an embryo include an egg in the process of fertilization, and, for this purpose, fertilization is not complete until the appearance of a two-cell zygote (HFEA 1990). PLA argued that because in the case of an embryo created by SCNT, fertilization was neither complete nor in progress, such an entity could not possibly be covered by the 1990 statute. The government urged the court to recognize that Parliament had wanted to craft a comprehensive regulation for the entire field of assisted reproduction and embryo research, and invited the judge to amend the original definition to "a live human embryo where [if it is produced by fertilization] fertilization is complete." But Judge Crane was not prepared to undertake an "impermissible rewriting and extension of the definition," concluding instead that an organism created by a procedure that had until recently only been imagined required an explicit ontological reframing by Parliament. Crane came to this conclusion after reading the "experts' illuminating statements," including the definition supplied by Dolly's "father," Ian Wilmut. Wilmut's characterization of the SCNT organism as a "reconstructed embryo" provided the judge with tangible, scientifically certified evidence that this new thing occupied an ontological limbo:

It may well be that responsible researchers would treat the organism as subject to control from an earlier stage, by analogy with s 1(1)(b) [the egg in the process of fertilization]. However, they would not be obliged to do so unless it became settled scientific practice to refer to the organism as an "embryo" before the two-cell stage. The very fact that an area of doubt would result is in itself an argument against the defendant's construction of s 1. (*Quintavalle v. Secretary of State for Health* [2001])

Judge Crane's handling of the case highlights the conflicts regarding the production of knowledge about living things. On the one hand, the court defers to science: the SCNT organism cannot be assimilated to an IVF embryo because the scientific practice is still unsettled, and eminent practitioners talk about "reconstructed embryos" rather than embryos tout court. The implication is that once the practice becomes settled, the courts no longer have doubts. On the other hand, this reasoning implies that it is ultimately the courts that certify when scientific practices have become sufficiently settled.

The judgment in favor of the PLA was reversed by the Court of Appeal, and the reversal was later confirmed by the House of Lords in *Quintavalle v. Secretary of State for Health*, House of Lords (2003). Lord Phillips of the appellate court reasoned that extending the definition of the 1990 Act to SCNT embryos was "plainly necessary to give effect to parliamentary intention," because it was clearly Parliament's purpose in 1990 to achieve a comprehensive regulation of experiments in human reproduction and embryology.

A close look at the decision of the House of Lords in 2003 shows the work of coproduction, through which judges defined the SCNT organism as an embryo. Lord Bingham first emphasized that "courts cannot fill gaps by asking what Parliament would have done in the current case" if Parliament had not in fact considered such cases at all. Their task "within the permissible bounds of interpretation, is to give effect to the parliament's purpose" (Quintavalle v. Secretary of State for Health, House of Lords [2003], paras. 10 and 8). Put concretely: "If Parliament, however long ago, passed an Act applicable to dogs, it could not properly be interpreted to apply to cats; but it could properly be held to apply to animals which were not regarded as dogs when the Act was passed but are so regarded now" (Quintavalle v. Secretary of State for Health, House of Lords [2003], para. 9, and Jasanoff 2005, 200). Here the "empirical judicial eye" (Jasanoff 2002, 37–69) claims for itself the capacity to sort out differences among and assign identities to changing biological entities. Asking whether the embryos created by SCNT "fall within the same genus of facts" as the embryos that Parliament had originally regulated, Lord Bingham answered with a plain yes:

An embryo created by in vitro fertilisation and one created by CNR are very similar organisms. The difference between them as organisms is that the CNR embryo, if allowed to develop, will grow into a clone of the donor of the replacement nucleus which the embryo produced by fertilisation will not. But this is a difference which plainly points towards the need for regulation, not against it. (*Quintavalle v. Secretary of State for Health*, House of Lords [2003], para. 15)

Thus, the matter of finding similarities and differences among organisms is intertwined with the need for regulation. They are two sides of the same coin. There is no appeal to science to deliver binding distinctions between IVF and SCNT embryos. Rather, the Lords affirmed their overall similarity in order to effectuate their reading of parliamentary intent. In short, one arrives at biological facts through the constraining filter of law. This is a coproductionist move: the overarching purpose of regulating a field of scientific inquiry ultimately carves out spaces in which newly contemplated entities can exist unproblematically.

The decision of the House of Lords has been criticized as an example of "paths of incremental policy-making inexorably progressing towards weaker restrictions" (Brownsord 2005). The argument is that because British case law proceeds in a piecemeal fashion, and because it uses institutionalized regulation (in this case, the HFEA), British policy is inherently prone to "incremental adaptation to practices that conflict with the need to protect human dignity" (Brownsord 2005). Those who voiced these concerns clearly think that there should be strong, time-tight barriers against certain scientific developments, and that the British system is incapable of erecting them. What appears problematic in this criticism is that it questions the very interplay of the scientific and the social that coproductionist scholarship has made explicit. The journey of the SCNT organisms through the British judicial and regulatory systems shows us the moves by which the United Kingdom has institutionalized coproduction. The active scene of British molecular embryology, the HFEA, and the British legal system form the corners of an extremely stable, coproductionist triangle, as the cloning controversy illustrates. The law orders an area of human activity according to a set of assumptions about the nature of parenthood, family, and reproductive rights, and this preliminary ontology is then mapped onto new artifacts coming from science. It is "well ordered science," not in Kitcher's (2001, 117–135) sense of deriving from ideal deliberators, but rather because the existing order from which it flows is so robust that new entities are actively prevented from creating ambiguity.

But why do the fact-finding practices of these two systems—law and the biosciences—seem to work so harmoniously in Britain? As Jasanoff (2005) has discussed in detail, British political culture and civic epistemology incorporate a preference for empirical demonstrations that converge with fundamental features of the scientific process. Both engage in practices of piecemeal fact finding with remarkably similar modalities. In the early years of biotechnology, the epistemologies of these two systems became particularly compatible with one another.

To clarify this point, let us return to Lord Bingham's point about cats, dogs, and in-betweens. In reminding us that the courts frequently determine whether modern inventions fall within old legal definitions, Bingham goes back to the case of *Grant v. Southwestern and County Properties Ltd* (1975), in which "Walton J had to decide whether a tape recording fell within the expression 'document' in the Rules of the Supreme Court." The judge concluded that the tape recording was a document, reasoning that "the furnishing of information had been treated as one of the main functions of a document." This functionalist turn says that a thing is defined by what it does. A similar turn has been observed in the life sciences. As Hans-Jörg Rheinberger (1997) points out, one of the distinguishing features of modern biology is that the object of investigation—the living thing—has also become the space, both physical and conceptual, in which experimentation takes place. Studying living matter means to modify it and ask questions about it from within. We discover things by (un)making them, and in this we can hardly succeed unless we ask what they do. Identities of biological things are defined by the actions that can be attributed to them: a notable example is the gene, whose physical boundaries are so difficult to pin down that its definition is almost entirely based on an understanding of its action. As Sarah Franklin notes, the emerging scientific discourse on stem cells posits "models of human life in which who we are, and what we are made up of, can be extracted and utilized in ways that are not only about the reuse of existing parts, but their redefinition. The redefinition of the human as a quantity of cells with different qualities is then further elaborated in terms of the ability to break down cellular capacities into specific functions, and to redesign them" (Franklin 2004, 63).

For a purposeful and inventive system like biotechnology, the common law framework that works by similarly reinscribing earlier assumptions is an ideal instrument for ordering human experience and human knowledge. Indeed, one of the strongest arguments to support this reading comes from a passage quoted by Lord Steyn in the HL judgment presented previously: "But it is one of the surest indexes of a mature and developed jurisprudence not to make a fortress out of the dictionary; but to remember that statutes always have some purpose or object to accomplish, whose sympathetic and imaginative discovery is the surest guide to their meaning" (*Cabell v. Markham* [1945]). So viewed, the law also discovers things, and the true meaning of a statute is worked out along the way, through the imaginative, empirically trained eyes of the courts. This orientation of the law as the lens through which the past can be made sense of in the present closely mirrors the epistemology of modern biology. As Jacob once remarked, experimental systems are a "machinery for the production of the future"—just as common law decisions are.

Italy: Artificial Clones, Natural Embryos

While the Donaldson Report was igniting the cloning controversy in the United Kingdom, the same controversy was being "brilliantly overcome" in Italy. With these words the Dulbecco Commission (DC), a body of experts appointed by former Health Minister Umberto Veronesi to advise the Italian government and Parliament on stem cells and cloning, described its work of drafting policy recommendations. A synopsis accompanying the press release of the Commission report on December 28, 2000 states: "[The novelty] that brilliantly overcomes the ethical quandaries raised by the Donaldson Report, lies in the adoption of the SCNT technology in order to obtain, while preventing the formation of the embryo, stem cells which should be immediately differentiated towards desired cellular lineages" (Synopsis of the Dulbecco Commission Report 2000). The report posited by definition that SCNT does not lead to the creation of an embryo but of something else posing no ethical quandaries.

To my knowledge, this is the first example of an advisory body explicitly constructing clones as not embryos. But there are other reasons for looking at this particular example of bioethical deliberation. Nearly two years after the DC issued its report, the argument that a clone was not like an embryo was used, as we have seen, to attack the British regulatory framework on embryo research. Thus, the juxtaposition of these two regulatory encounters with cloning shows how the same alleged biological fact can be construed in completely different ways to fit with preexisting normative commitments. The PLA lobby and the DC obviously wanted to achieve opposite results, but both were able to draw the necessary legitimation from the same piece of scientific reasoning.

The Italian debate on assisted reproduction and embryo research had been developing in a highly polarized form between Catholics and lay-persons. But, in contrast to Britain, Italy had not achieved any closure comparable to the HFEA of 1990 and remained until February 2004 without any comprehensive legislation. In this legislative void, examples of radical social experimentation based on novel technology—especially sur-rogate motherhood and postmenopausal pregnancy—hit the headlines and provoked calls for an "end to the reproductive Wild West" (see Metzler, chapter 5, this volume). Yet the Italian civil law system, equipped to issue either comprehensive legislation or none at all, did not allow a piecemeal confrontation with reproductive technologies.

Minister Veronesi summed up the extent of the political polarization, and by implication the mission of the DC, in an interview that accompanied the release of the commission report: "Ours is a Catholic country. And when there is strong opposition for religious reasons, it is useless to carry on fights over principles. It is better to find, if possible, intermediate solutions" (2000, author's translation).

The DC was composed of twenty-six members: seventeen life scientists and physicians, four philosophers, three theologians (including a cardinal), one judge, and one politician. The minister asked the commission to consider four questions: the therapeutic potential of stem cells, the time within which this potential could be achieved, the range of diseases that could be affected by such developments, and which source of stem cells (embryonic, adult, fetal, and cord blood) was scientifically and ethically preferable. These questions were immediately criticized for the consequentialist logic that placed ethical and social concerns downstream of scientific arguments, within a framework apparently presupposing the therapeutic value of this research (Maio 2001, 299–309).

Predictably, the commission split over the moral permissibility of deriving embryonic stem (ES) cells from surplus IVF human embryos. The secular majority appealed to the principle of beneficence, while the Catholic minority saw the practice as an unacceptable instrumentalization of human life. Equally unsurprising was the commission's unanimous endorsement of research on adult, fetal, and cord blood stem cells.

The surprise came when the commission unanimously agreed that human SCNT presented no special ethical quandaries and should be energetically endorsed. How was this remarkable consensus achieved? The chapter dedicated to SCNT states:

The term "therapeutic cloning" to indicate SCNT is clearly inappropriate. In fact, an enucleated oocyte reconstructed with an adult somatic cell nucleus cannot be considered as a classical zygote, because it does not derive from the union of two gametes. This is proven by the fact that such a reconstructed oocyte does not develop spontaneously into an embryo, and this happens only following artificial stimulations that force it to develop into a blastocyst. Only few of these blastocysts possess the effective capacity of forming an embryo, and hence a fetus, once transferred into the uterus. . . . . Finally, the oocyte reconstructed with a somatic cell nucleus is much more similar to a potential form of asexual cellular expansion of the patient, in analogy to what is currently practiced when skin biopsies are amplified in vitro to produce artificial skin for the treatment of major burns. (Dulbecco Commission Report 2000; author's translation)

Consistent with these premises, the DC concluded that SCNT, along with adult and cord blood stem cells, did not present insurmountable ethical problems "so long as the SCNT product does not develop into a human embryo."

Boldly reclassifying the product of SCNT from cloned embryo to unproblematic multipotent cells, the DC treated it merely as an extension of a person's body. This definition was emphasized throughout the text as the commission's own conceptual contribution to the emerging field of stem cell research and regenerative medicine. Following suit, all major Italian newspaper presented SCNT as the "Italian way to stem cells," conflating cloning, national pride, and the desire for status on the international research scene into a single powerful sociotechnical imaginary (Jasanoff and Kim 2009).

The commission's main argument—namely, that the reconstructed egg produced via SCNT could be led to form differentiated cells without ever intersecting the development of a proper embryo—was criticized, especially in the Catholic press (Maio 2001, 299–309). Although the reprogramming of cell fate without having to go through embryonic development was the ultimate goal of many scientists, none had yet demonstrated its feasibility. But most relevant for our analysis is the way in which this political body reframed the object of cloning in such a way as to render irrelevant any question of moral status or rights. Bypassing the blocked ontological deliberation about the status of the embryo, the DC carved a niche of legitimacy for SCNT clones by framing them as artificial products of human ingenuity, both in the way they were created (nuclear transfer as opposed to fertilization) and in the way they were to be employed (their conversion directly into cells without undergoing embryonic development).

Interestingly, the Italian Law 40 on assisted reproduction and embryo research, passed in 2004 by a Catholic majority amid heated contestation, adopted the DC's discourse, albeit with opposite results (see also Metzler, chapter 5, this volume). At first sight, the two texts could not be more different. Whereas the DC had advocated SCNT, Italy's Law 40 strictly regulated both assisted reproduction and embryo research. But both the DC and the Italian Parliament resorted to nature as a source of political legitimation for their norms. The first article of Law 40 protects all subjects taking part in the process of assisted reproduction, including the zygote. The fifth article licenses assisted reproduction solely as a technological alternative for building the natural family: assisted reproduction is allowed for "all couples of different sex, older than 18 years, married or living together, both in fertile age and both living." The law also forbids cloning and any kind of ES cell derivation from IVF embryos as well as preimplantation genetic diagnosis. And to prevent the appearance of "supernumerary embryos," the law prescribes that all embryos produced in an IVF cycle (up to a maximum of three) be simultaneously implanted.

With respect to embryos from women who refuse implantation, the guidelines specify that the "*in vitro* culture of the embryo needs to be continued till its extinction." *Extinction* is a loaded—and rare—word in Italian. It conveys the affective and relational aspect of death, as in the expression "caro estinto" (dear extinct one) used to honor one's dead during funerals or memorials. It is hard to envision how that concept of extinction applies to an IVF embryo in an incubator, but the spirit of the law is precisely to invest a technological process with natural categories. The production of the in vitro embryo, which must serve only natural purposes, is understood as the prosthetic extension of normal life processes, and the practice of terminating a cell culture is therefore infused with the ritual meaning that accompanies death. In a word, extinction.

Thus, the DC and Italian Law 40 both construct their problem framing around the divide between natural and artificial. The DC uses this divide to carve out legitimate spaces for new entities that can be brought into existence by virtue of their not being natural. The Italian lawmakers use the same divide to simultaneously block off social and scientific experimentation. In the United Kingdom, both during the parliamentary debate of the 1990s and the cloning controversy of the early 2000s, "nature" dispersed through the domains of social negotiation to the point where the difference between fertilization and nuclear transfer became legally irrelevant. In Italy, by contrast, once the natural embryo was constructed as a bearer of rights on a par with other citizens, clones could come into being only as artificial products whose very dissimilarity from nature was also the source of their legitimacy.

Thus, in its relation to biotechnological change, Italian political culture appears to be structured around two rigid dichotomies: the secular/religious and the natural/artificial, which refer to the two main sources of certified authority (science and the Catholic Church) and the historical divide along which they strive to maintain their separate spaces. The designation of experts in bioethical debates reinforces this point. Neither the mandate nor the composition of the DC included, for example, any representative from groups such as patients or consumers not usually associated with certified expertise.

The United States: Political Clones

On December 3, 2004, William Hurlbut, a physician from Stanford, and member of the U.S. President's Council on Bioethics, presented to the council a "technological solution" to solve the "moral impasse" on SCNT, claiming that "a purely political solution will leave our country bitterly divided, eroding the social support and sense of noble purpose that is essential for the public funding of biomedical science" (Hurlbut 2004). His diagnosis of a bitterly divided country was correct, as the controversy around embryonic stem cell research and cloning had reignited the tension over early human life that dated at least as far back as *Roe v. Wade* (1973). Hurlbut's proposal, referred to as Altered Nuclear Transfer (ANT), was straightforward enough. If, prior to the transfer of the somatic nucleus into the enucleated egg, we could inactivate in the somatic nucleus a gene known to be essential for development, this novel biological entity would after transfer "lack the attributes and capacities of a human embryo," being "biologically and morally more akin to the partial organic potential of a tissue or cell culture" (U.S. President's Council on Bioethics, 2005, 36–37). The biological modification should be radical enough to change the entity's moral status; yet, by definition, it should still enable this entity to arrive at the developmental stage (the blastocyst) required for the derivation of ES cells. The moral disarray of developmental options opened up by SCNT is domesticated in this case through the constitution of a new artifact that is at once a product of moral and biotechnological ingenuity. Moral ingenuity emanates from the deeply held commitment of one part of American society to regard life as ontologically distinct and morally valuable from the moment of fertilization. Biotechnological ingenuity serves in turn that very commitment, recruiting the toolkit of genetic engineering to create a form of life that bypasses, by design, the ontological and moral fences that have defined life's beginning. The result is a form of life that, by the moral definition that brings it into being, never began; a form of "growth without life," as Hurlbut, George, and Grompe (2006, 42–50) put it. For this reason, I argue that this domestication is a constitutional moment. It is the first explicit instance of inscribing moral status into a living entity by means of genetic engineering (Testa 2008).

In the first half of this coproductionist cycle, scientific understandings and experimental practices from systems biology and mouse embryology are taken up to fit into an existing framework of values and norms. Thus, Hurlbut (2004) distinguished the human embryo as a whole entity, in which "this principle of organismal unity is an engaged and effective potential-in-process," from mere parts of the whole, which he called "subsystems with partial trajectories of development . . . that ultimately fail to rise to the level of coordinated coherence of a living organism." The language of system biology, with its emphasis on the "whole" and the "integration of the parts," was thereby imported into a preexisting notion of human moral worth that starts at fertilization. The coalescence of scientific arguments and moral convictions results in the view of the human embryo as containing from the moment of fertilization not "an abstract or hypothetical potential in the sense of mere possibility, [but rather] an engaged and effective potential-in-progress, an activated dynamic of development in the direction of human fullness of being" (Hurlbut 2004). This articulation is reminiscent of the position expressed by the Catholic members of the Dulbecco Commission, who condemned the derivation of ES cells from supernumerary embryos on the grounds that the human embryo is a "human being with potential, and not a potential human being" (Maio 2001, 299–309).

Yet if the holistic framing of ANT aimed at preventing the formation of embryos as wholes, its implementation required tinkering with them in their constituent parts, mining the reductionism of mouse embryology for both concepts and tools. Specifically, Hurlbut presented Cdx2 as an ideal candidate gene to be inactivated prior to the SCNT, because studies in the mouse had shown that it was required for the formation of the placenta and hence for implantation into the uterus. Two technologies of inactivation that have transformed molecular biology in the last two decades stood available: conditional mutagenesis and RNA interference, whose combination enables one to dampen the activity of a given gene in a reversible manner. Inactivation of Cdx2 seemed then radical enough to disable organismal unity, yet still compatible with the development of ESCs from the ANT product; furthermore, the reversibility of Cdx2 inactivation meant that its function could easily be restored if it turned out to be essential in one of the ANT-derived lineages.

The second half of this coproductionist cycle followed straightforwardly enough. On October 16, 2005, *Nature* published the experiment originally proposed by Hurlbut and carried out by the team of Rudolf Jaenisch at MIT (Meissner and Jaenisch 2006a, 212–215). The coalescence of moral commitments with epistemic assumptions had literally constituted a new life form that bore two stamps of authority: the political authority of the inspiring body (the U.S. President's Council on Bioethics) and the scientific authority of the executing lab and the high-ranking journal.

The extent of this conflation between the political and the scientific, the normative and the epistemic, is exemplified in a discussion in the *New England Journal of Medicine* among scientists respectively endorsing or opposing ANT (Meissner and Jaenisch 2006b; Nowotny and Testa 2009, 97–98). One controversy dealt with the technology used to inactivate Cdx2. Because ANT works not by physically removing the gene from the cell but by dampening its expression, the question arose as to how one could be certain that one had lowered the levels of Cdx2 enough, and hence disabled organismic potential enough, in order for the entity created through ANT to be morally unproblematic—a thing rather than a precursor to the human. The toolkit of molecular biology was once again mined for a technological solution, this time in the guise of Green Fluorescent Protein (GFP). This protein emits a green light when subjected to blue light and has become the most widely used tool for live imaging in cells or indeed entire animals. The solution to the problem was then to let GFP act as a readout of the levels of Cdx2: the brighter the GFP signal, the greater the efficiency of the Ribonucleic Acid Interference (RNAi) technology, the lower the residual levels of Cdx2, and hence the more morally unproblematic the resulting entity. Long used as a reporter of life functions, GFP was now figuring as a reporter of moral worth as well.

Perhaps the best description of this coproduction was inadvertently given by Michael Gazzaniga, also a member of the U.S. President's Council on Bioethics and a strong supporter of embryonic stem cell and research cloning. "Normally," he said, "we generate a word to describe a biological phenomenon, and here we seem to be tinkering with biological phenomenon to have it fit the meaning of a word" (Gazzaniga 2004). *Logos* comes before *bios* in this case, and that is precisely the point. The embryo's moral status is cast as a question about its potential to achieve maturity. So translated, that potential is assessed within the conceptual and practical framework of molecular genetics; hence, the moral status of the embryo ultimately becomes predicated on the technical reversibility of conditional mutagenesis, as a property that can be switched off and on biologically, within the experimental paradigm of molecular switches.

What comes into relief from this dense encounter of science and politics is a distinct U.S. modality of pursuing social order and the stability of moral values in the face of biotechnological change. The constitutional dimension of this modality is captured well in the figure of the editorial that accompanied the *Nature* article on ANT (Weissman 2006, 145–147). The figure depicts the range of developmental forms at the edge of life generated through biotechnology in recent years. The SCNT embryo faces two possible destinies in the shape of two arrows: the first is implantation in uterus, the second is transformation into ESCs. The label "ban?" sits over the first arrow, hinting at the law as a way of preventing reproductive cloning. For the ANT product, the arrow pointing toward implantation is crossed out (no Cdx2 and hence no need for a ban), and the only remaining option is the derivation of ESCs. This illustration entails then a direct comparison between law and genetic engineering as two technologies of moral order. In the case of standard SCNT, the law should come in with a ban to block off the reproductive implications of these entities; the slope might be slippery, but we "can stop our descent . . . at any point when we wish to do so and the way of stopping ourselves descending into unknown horrors is by legislation" (Warnock 1988, cited in Jasanoff 2005, 156). ANT promises to achieve the same result but through technological means, and its uptake by politicians and scientists implies that in the age of biotechnological kind making, genetic engineering becomes, alongside the law, a significant tool of social ordering. By in effect intertextualizing new life forms with moral rationality, this proposal represents an example of constitutional experimentation in an age of biotechnological change.

Conclusions

In all three cases of the making of clones discussed in this chapter, the clones were actively coproduced in the scientific and social domains. For one thing, the question of what a clone is was addressed together with the broader question of what one should do about it. The epistemic, the ontological and the normative operated, from the very start, as three aspects of the same undertaking, pointing to a first, self-evident level of coproduction. This is especially obvious in the British and the Italian cases: both the Law Lords and the members of the Dulbecco Commission framed the nature of the clone within the explicit mandate to draft the legislation about it. But equally in the American case, the Hurlbut proposal was designed to overcome a political and moral impasse through a biotechnological fix.

Beneath this commonality, however, we also detect differences in the ways in which coproduction unfolded in each national setting. The main difference may be crudely summarized as follows: the HL and the DC were talking about clones already produced in the lab, while the U.S. President's Council on Bioethics was talking about something that would be made only following its suggestions. But this observation points to a more important difference. In the United Kingdom and Italy, the dispute was about where to draw a boundary, and about where to place new kinds of living matter: among the known and natural or among the new and artificial. As we have seen, the answers differed sharply. In contrast, when Hurlbut proposed ANT to the Council on Bioethics, it appeared that the boundary between natural and artificial was already in place. Of course, the scientists and ethicists practicing ANT also needed to work their way through an ontological characterization of the resulting entities. But this deliberation was undertaken precisely in order to prove that the new entity does not in any way undermine boundaries that are already settled. These two patterns of coproduction support with empirical evidence the interpretive framework that distinguishes between constitutive and interactional coproduction (Jasanoff 2004, 13–45).

As instances of constitutive coproduction, both the British and the Italian cases deal with the emergence of a coproduced object. The main preoccupation in both national contexts regarded how to name and classify it. The HL and the DC, respectively, were called upon to map—or remap—the bodily derivatives and cellular potential inhabiting and surrounding persons such as Tessa Wick. To what extent were the new biological objects akin to other bodily derivatives already accommodated within law and ethics? Did Tessa Wick have jurisdiction over this new combination of her cells with advanced technology, or was the resulting artifact sufficiently "other" so as to deserve a public reappraisal of its nature and its moral implications? As we have seen, both the HL and the members of the DC endorsed the creation of clones as research objects, but they arrived at that conclusion by different means.

In the United Kingdom, conceptual resources were already available to order and classify the product of SCNT. The debate of the late 1980s on assisted reproduction and embryo research, leading to the HFEA of 1990, had framed the in vitro embryo during its first fourteen days of existence as a legitimate object of scientific research. The process had succeeded largely by constructing the fourteen-day limit as a biological boundary that was out there to be observed, and as a demarcation available to common sense and hence also to communal moral sensibilities. Guided by expert knowledge and reasoning, Parliament and the British public learned to see, as if it were self-evident, that the pre-fourteen-day embryo had a biological nature, and hence a space of rights, distinct from the embryo itself (Jasanoff 2005, 147–160).

In 2003, this solid consensus, made even stronger by the intervening thirteen years of implementation and practice, could be easily mapped onto the newcomer—the product of SCNT. The potential fuzziness brought about by a new way of making life was resolved and repositioned within preexisting categories. And in this case, perhaps even more than in the HFEA deliberations, judicial reasoning highlights the centrality of common sense to the functioning of British law and politics, the trust in the power of the empirical observation of reality that legitimates public policy. Nature emerges as a remarkably stable and consensual place, a repository of entities and demarcations available to direct public scrutiny. But, and this is the point, stability does not come from the outside; it is part and parcel of the political culture (Jasanoff 2005). Through the pervasive, if implicit, appeal to common sense and common vision, all the work required for aligning scientific and political distinctions dissolves—as if those distinctions were truly out there, as if they could be perceived, un-aided, by the "eye of everyman" (Jasanoff 1998, 713–740).

While speaking about the clones, the Law Lords necessarily spoke about themselves and their role in British public life. Framing the clones within the statutory framework of the HFEA meant at the same time deciding the limits within which judges are allowed to interpret the original intention of Parliament, and the legitimate techniques for doing so. The practices of molecular biology, always open to future surprises and aimed at extracting and transforming living functions, were legitimized and reinforced by a legal system that did not pose ontological questions about the essence of new beings but only functional questions aimed at how to regulate them.

The discussion in the Dulbecco Commission was also about new bound-aries to be drawn in an uncharted space, and it is thus another example of constitutive coproduction. But in Italy there was no prior consensus on the ontology of the in vitro embryo. Though assisted reproduction had flourished in a regulatory vacuum, the backdrop was a tacit divide between Catholics and laypersons on the moral status of the embryo, a di-vide eventually enshrined in a final report that split over the permissibility of conducting research on supernumerary IVF embryos. When cloning presented the possibility of blurring developmental trajectories, this new way of making life entered a discourse in which no compromise seemed possible. Thus, the same properties of clones that had seemed irrelevant to the British courts (asexual origin, artificiality, etc.) were drawn upon to make exactly the opposite argument. Sex (the production of embryos through fertilization) was aligned with the category of the natural in a construction that carved for the asexual clones a niche of artifactual, and hence legitimate, existence as an object of research. In doing so, the DC reinforced a tacit understanding of the embryo as a legitimate subject of rights and of the "natural" as a relevant category for organizing scientific and social practices. The construction of the clone as an extension of a patient was seen as a brilliant compromise that could overcome the divide between Catholics and laypersons. Yet it served at the same time to strengthen this divide and confirm it as a meaningful, even inevitable, articulation of Italian politics, which resurfaced in the contentious politics of Law 40 (see Metzler, chapter 5, this volume).

Finally, the Hurlbut proposal and its scientific implementation in the United States displayed features of both varieties of coproduction, the interactional and the constitutive. SCNT products were positioned in the United States within the same contested domain as IVF embryos and aborted fetuses. Against this backdrop, science and politics tried to solve the conflict by accepting and reinforcing each other's boundaries, the hallmark of interactional coproduction. The means chosen for this purpose synchronized well with America's predilection for maintaining a strict divide between epistemic and normative domains, and for finding technological solutions to difficult or contentious moral problems (Jasanoff 2005). In the process, however, these two distinct domains (epistemic versus normative, gene switches versus moral values) coalesced into a single object that explicitly combined the stamps of scientific authority and political legitimacy. It literally owed its existence to criteria that became scientific only within a template of moral assumptions, and that became political only within a template of scientific options. In this sense, the interactional coproduction that reinforced preexisting boundaries evolved into an instance of frankly constitutive coproduction. Moral and religious commitments about the beginning of human life and the source of its dignity, political commitments about the necessity of consensus, and a social understanding of science as a public enterprise (that therefore requires consensus) fell together with the molecular explanation of life and the most sophisticated tools for its genetic engineering. We witness, quite literally, the mutual and constitutive reinforcement of state making and science making.

Precisely through this conflation of the political and the scientific, likely prophetic of future developments, ANT vividly highlights the power and resolution of the coproductive lens. Facing the task of finding anchor points in order to reveal interactions between the blurred, overlapping, and mutually shaping domains of science and society, the challenge for the analyst is to treat these categories as sufficiently distinct as to make them useful while leaving them sufficiently porous as to question what counts as scientific or social in each instance. In requiring symmetry of analysis and denying primacy to either the scientific or the social, the coproductionist framework suits this task particularly well. It allows us to expose ANT not as the conflation of two "factually" distinct domains, but rather as the conflation of two domains that are combined together precisely insofar as they are treated as distinct, precisely insofar as their distinction is considered an important resource in American public discourse. Hence it allows us to see how the final outcome (the ANT artifact) denies that distinction at the level of the flesh—by translating values into gene switches—while upholding it at the level of the polity as a meaningful resource for the solution of ethical dilemmas.

In the end, the juxtaposition of these three cases highlights how political cultures (their historical constraints, their discursive resources, and their ways of distributing and recognizing expertise) are integral to the development of technoscientific objects. In reverse, we see how the encounter with scientific objects, and the need to articulate their public meanings, are "moments of truth" in which political cultures affirm, discover, or indeed reinvent the sources of their legitimation. The allegedly same object—the cloned cell—was framed in different ways and granted a different ontological and legal status in the three cultures we examined. But comparison reveals that this diverse ordering did not result from a confrontation between a predefined object and equally predefined legal and ethical principles that could either accept or reject it. Preexisting instead were institutional features and conceptual resources, on both the technoscientific and the legal-political sides. And if we now return to Tessa Wick's imagined domain of jurisdiction over the developmental potential of her cells, we can finally make sense of how those features and resources were aligned, in three different political cultures, on the basis of different conceptions of what counts as natural and good, and of how we come to recognize either. These different conceptions guided then the emergence of three constitutional dispensations that differentially enabled, or disabled, the public circulation of clones.

References

Brownsord, Ronald. 2005. Regulating Stem Cell Research in the UK: Filling in the Cultural Background. In *Grenzüberschreitungen. Ethische, Politische und Religiöse Aspekte der Stammzellenforschung*, ed. W. Bender, C. Hauskeller, and A. Manzei, 413–433. Muenster: Agenda Verlag.

Donaldson Report. 2000. "Department of Health. Stem cell research: Medical progress with responsibility." Retrieved from <http://www.dh.gov.uk/en/Publications andstatistics/Publications/PublicationsPolicyAndGuidance/DH_4065084>.

Dulbecco Commission Report. 2000. Retrieved from <http://www.lucacoscioni .it/node/2349>.

Franklin, Sarah. 2004. Stem Cells R Us. In Global Assemblages: Technology, Politics, and Ethics as Anthropological Problems, ed. A. Ong and S. J. Collier, 59–78. Malden, MA: Wiley-Blackwell.

Gazzaniga, Michael. 2004. Transcript of the U.S. President's Council on Bioethics session "Seeking Morally Unproblematic Sources of Human Embryonic Stem Cells." Retrieved from <http://bioethics.georgetown.edu/pcbe/transcripts/dec04 /session6.html>.

HFEA. 1990. "Human Fertilisation and Embryology Act 1990." Retrieved from <http://www.legislation.gov.uk/ukpga/1990/37/section/1/enacted>.

Hurlbut, William. 2004. "Altered Nuclear Transfer as a Morally Acceptable Means for the Procurement of Human Embryonic Stem Cells." Commissioned working paper. Retrieved from <http://bioethics.georgetown.edu/pcbe/background/hurlbut .html>.

Hurlbut, William. 2005. Altered Nuclear Transfer as a Morally Acceptable Means for the Procurement of Human Embryonic Stem Cells. *National Catholic Bioethics Quarterly* 5 (1): 149.

Hurlbut, William, Robert George, and Markus Grompe. 2006. Seeking Consensus: A Clarification and Defense of Altered Nuclear Transfer. *Hastings Center Report* 36 (5): 42–50.

Jasanoff, Sheila. 1998. The Eye of Everyman: Witnessing DNA in the Simpson Trial. *Social Studies of Science* 28 (5–6): 713–740.

Jasanoff, Sheila. 2002. Science and the Statistical Victim: Modernizing Knowledge in Breast Implant Litigation. *Social Studies of Science* 32 (1): 37–69.

Jasanoff, Sheila. 2004. Ordering Knowledge, Ordering Society. In *States of Knowledge: The Coproduction of Science and Social Order*, ed. Sheila Jasanoff, 1–45. London: Routledge.

Jasanoff, Sheila. 2005. *Designs on Nature: Science and Democracy in Europe and the United States*. Princeton: Princeton University Press.

Jasanoff, Sheila and Kim, Sang-Hyung. 2009. Containing the Atom: Sociotechnical Imaginaries and Nuclear Power in the United States and South Korea. *Minerva: A Review of Science, Learning & Policy* 47 (2): 119–146.

Kitcher, Philip. 2001. *Science, Truth, and Democracy*. New York: Oxford University Press.

Maio, Giovanni. 2001. Die ethische Diskussion um embryonale Stammzellen aus internationaler Sicht—das Beispiel Italiens. *Zeitschrift für medizinische Ethik* 3:299–309.

Maienschein, Jane. 2003. *Whose View of Life? Embryos, Cloning and Stem Cells*. Cambridge, Mass.: Harvard University Press.

Marcus Aurelius. 2005. *Meditations*, trans. George Long. Originally written 170–180 A.D. Digireads.

Meissner, Alexander, and Rudolf Jaenisch. 2006a. Generation of Nuclear Transfer-Derived Pluripotent ES Cells from Cloned Cdx2-Deficient Blastocysts. *Nature* 439 (7073): 212–215.

Meissner, Alexander, and Rudolf Jaenisch. 2006b. Politically Correct Human Embryonic Stem Cells? *New England Journal of Medicine* 354: 1208–1209.

Nowotny, Helga, and Giuseppe Testa. 2009. *Die gläsernen Gene. Die Erfindung des Individuums im molekularen Zeitalter*. Frankfurt am Mein: Suhrkamp.

Rabinow, Paul. 1996. *Essays on the Anthropology of Reason*. Princeton: Princeton University Press.

Rheinberger, Hans Joerg. 1997. *Towards a History of Epistemic Things*. Stanford: Stanford University Press.

Synopsis of the Dulbecco Commission Report. 2000. "Sintesi della relazione della 'Commissione Dulbecco.'" Retrieved from <http://fondazionebassetti.org/0due/threads/03commissione-dulbecco.htm#documenti>.

Testa, Giuseppe. 2008. Stem Cells through Stem Beliefs: The Co-production of Biotechnological Pluralism. *Science as Culture* 17 (4): 435–448.

U.S. President's Council on Bioethics. 2005. "Alternative Sources of Human Pluripotent Stem Cells." Retrieved from <http://bioethics.georgetown.edu/pcbe/reports/white_paper/index.html >.

Veronesi, Umberto. 2000. Interview in *La Repubblica* (December 28).

Weissman, Irving. 2006. Medicine: Politic Stem Cells. *Nature* 439 (7073): 145–147.

Wilmut, Ian. 1997. Viable Offspring Derived from Fetal and Adult Mammalian Cells. *Nature* 285 (6619): 810–813.

Cases Cited

Cabell. v Markham (1945) 148 F. 2d 737.

Quintavalle v. Secretary of State for Health (2001). 4 All ER 1013.

Quintavalle v. Secretary of State for Health, House of Lords (2003). 2All ER 113.

Roe v. Wade (1973). 410 US 113.

Grant v. Southwestern and County Properties Ltd (1975), chap. 185.

5

Between Church and State: Stem Cells, Embryos, and Citizens in Italian Politics

Ingrid Metzler

By the evening of June 13, 2005, Cardinal Camillo Ruini, then head of the Italian Bishops' Conference (*Conferenza Episcopale Italiana*), appeared deeply satisfied.[1] It was the evening following a national referendum in which the Italian electorate had voted on the relaxation of Italy's "norms in matter of medically assisted procreation" (*norme in materia di procreazione medicalmente assistita*) (Repubblica Italiana 2004). The results, as the cardinal explained on the evening news, had "exceeded all [his] expectations" (La Repubblica.it 2005). The referendum was the first test of the law that had moved Italy from the unregulated and permissive end of the regulatory landscape to its most restrictive edge. Italy's voting citizens had a chance to decide whether they wanted to modify the law in such a way as to remove many of the restrictions on human embryonic stem cell (hESC) research. Yet only a quarter of the electorate turned out to cast their votes in this referendum. This was precisely what the cardinal had sought to achieve when he had called on Italians to abstain from voting.

Once the polls closed, the Cardinal modestly declared that he had "only sought to carry out [his] duty as bishop" (qtd. in Politi 2005, 2). But before the referendum, he had chosen a rather worldly strategy of political mobilization (or, more accurately, *im*mobilization) to implement his pious mission. Although the Italian constitution enables abrogative referenda to repeal a law in whole or in part, such a referendum can only be valid if a minimum of 50 percent plus one of the Italian electorate casts a vote. Thus, in January 2005, when the Constitutional Court approved the embryo research referendum, the cardinal chose not to rely simply on the docile souls of his flock, but also on the inert bodies of those Italians who never cast their votes in such proceedings. Apparently, the cardinal deemed it more feasible to convince 50 percent of the Italian electorate to not participate than to trust the majority to choose the right answer. Ruini's strategy of promoting abstention (or electoral abstinence) was later embraced by other groups. And it proved successful. As 74.3 percent of voters deserted the polls, the referendum was invalid and Ruini's mission was accomplished. Though the Catholic hierarchy's invitations to sexual abstinence may often have gone unheard by their unmarried flocks, the Cardinal's call for political abstinence seems to have been not only heard but also gladly accepted by the Italian electorate.

The result was that Law 40 (*legge quaranta*), as the law is conventionally termed, continued to bar Italian scientists from "killing" Italian embryos for stem cell procurement but did not forbid them from engaging in this line of research altogether. Italian scientists are still allowed to conduct hESC research, provided that they draw on hESC lines that are imported from abroad and hence not derived from Italian embryos. As yet, the work of the barely half-dozen groups who engage in hESC research in Italy is tacitly tolerated but not publicly supported. All national funds are invested in research projects that draw on stem cells derived from animals, adult tissues, cord blood, or aborted fetuses. Scientists engaging in hESC research depend on private funding or funding from the European Union's Framework Programmes (Abbott 2009; Cattaneo et al. 2009). But why has Italy refrained from embracing a line of research that many other nations have seized upon? And why was Italy's abstinence contested? In this chapter, I attempt to make sense of these questions. I argue that Italian hESC politics emerged from the flip side of a larger bioconstitutional project in which the foundations of Italy's biopolitics were reordered through the "nationalization" of Italian embryos. Italian hESC politics provides a window through which this bioconstitutional project can be explored through the back door. Although the nationalization of Italian embryos restricted the material of Italian hESC research, it also gave Italian hESCs their particular meaning as signifiers of a battle for rights and liberties against an oppressive state and its Roman Catholic ally.

I start with a brief discussion of an early period of Italian hESC politics in which hESC research was unregulated and very much a nonissue.[2] This period ended in February 2004, when Italian legislators regulated the range of permissible practices in Italian fertility laboratories, establishing the "norms in the matter of medically assisted procreation" (Repubblica Italiana 2004). In this law, Parliament redefined the collective of Italian citizen subjects, redefining the status of IVF embryos as subjects with no duties but a set of legal rights (Filippini 2004; Hanafin 2006). Embryos that used to be under the jurisdiction of medical professionals and prospective parents were withdrawn from their control and put under the guardianship of the state—they were effectively "nationalized."

It was in this bioconstitutional moment that hESCs left their places of containment and moved to the center of Italy's political stage. Along with infertile women and couples, feminists, medical professionals and scientists, sick citizens complained that the legislators' care for the rights and due protection of Italy's newest citizens effectively infringed on *their* rights. The resulting fierce debates peaked in a national referendum in June 2005. Much of my article will focus on this period. I do not claim that this exhausts Italian hESC politics, or that it is particularly representative. Yet this period is instructive for understanding Italian hESC politics and the larger bioconstitutional project from which hESC politics took shape. This did not concern only whether Italy should endorse hESC research; hESCs were also entangled with more constitutional debates that put the very categories of the Italian polity at stake, opening questions such as who belonged to the moral community of Italian citizen subjects and what kind of state they should be subjects of. Italy's soul searching on how to come to terms with hESC research addressed questions about what life is and what it should be, about who is allowed to speak and act on it, and about the appropriate places of law and the state in answering these. To make sense of this debate, we have to understand an earlier bioconstitutional moment when Italy revised the rules of biopolitics that Law 40 had written into law. In the last part of this article, I therefore return to the period in which those rules were initially articulated: the controversies on assisted reproductive techniques in the two closing decades of the last century. I explore how Italy's sociopolitical actors tried to make sense of the new choices and responsibilities, and how they eventually resolved these controversies by nationalizing embryos and naturalizing parts of human biology.

An Issue of Lowest Interest

Human embryonic stem cells made their first public appearance in 1998 when James Thomson and colleagues reported to the readers of *Science* that they had successfully derived hESCs from early human IVF embryos (Thomson et al. 1998). These small entities quickly spread from the pages of scientific journals to newspaper headlines, engendered hopes of cures and therapies, and became part of policy makers' agendas in countries within and outside Europe (Jasanoff 2005a; Gottweis et al. 2009). Some states drafted regulations that attempted to redirect the cells' potentiality into enhancing the prospective health and wealth of their nations. Other states, such as Germany and the United States, were more reluctant, and they puzzled about how to reconcile the promises of hESCs with their constituencies' understandings of the due protection of human life. Yet as most Western states ventured into project of endorsing the cells' vitality in laboratories while seeking to tame their appearances in public moral discourse, in Italy the appearance of these new entities was first met with silence: they were neither embraced nor expelled from the national imaginary: they were simply not an integral part of it.

All forms of animal and human cloning were banned in Italy by ministerial ordinance in 1997 (Ministro della Sanità 1997).[3] Italian scientists were therefore enjoined from using cloning techniques to produce embryos for hESC procurement, which—as discussed by Testa (chapter 4, this volume)—was one path that scientists had mapped for producing personalized hESC lines.[4] But national law did not prohibit the procurement of stem cells from normally fertilized embryos. In 1985, Health Minister Costante Degan issued a circular barring public fertility centers from producing "embryos in excess"—that is, embryos not immediately implanted—and from using embryos for research (Ramjoué and Klöti 2003). It did not affect the practices of private fertility centers. However, remarkably for a Catholic country, the possible presence of hESCs in Italy's laboratories attracted little attention. News about hESC research was relegated to the science sections of Italian newspapers. The public imagination was captured not by stem cell stories from foreign countries but by "fertility" stories emerging from Italy's private reproductive clinics, where doctors were spawning strange kinship relations and postponing the reasonable age of motherhood. Practices in these clinics were regulated not by national laws or regulations but by soft laws and the moral sensibilities of medical professionals and prospective parents. Although some praised this "legal void" (Neresini 2000; Ramjoué and Klöti 2003) as a situation of "freedom of conscience," conservative commentators in particular damned it as Italy's "Wild West" (*Far West*) of reproduction (Cazzullo 2003; De Bac 2003).

While policy makers in other countries were sizing up the opportunities for hESC research, Italian policy makers were primarily dealing with reproductive issues. The final report and recommendations of the Dulbecco Commission, an expert commission on the risks and opportunities of stem cell research that Health Minister Umberto Veronesi established in August 2000, generated some perplexities when it was issued in December 2000 (see Testa, chapter 4, this volume). Yet the report itself was shelved when Health Minister Girolamo Sirchia succeeded Veronesi in the spring of 2001 (Ministero della Sanità 2000). At that point, hESCs were removed from Italy's political stage before they became an integral part of it. In the words of a key policy maker, they were an issue "of lowest interest" that might be debated by scientists but was of no public concern and therefore not in need of state actions.

This period came to an abrupt end in early 2004. On February 10 of that year, Italy's Chamber of Deputies—one of the two branches of the Italian Parliament—put an end to a difficult and drawn-out process of legislation by enacting Law 40 (Repubblica Italiana, 2004). In their attempt to redraw the boundaries of the Italian family through ordinary law, the legislators ended up with a bioconstitutional moment. They expanded the legal boundaries of the community of citizen subjects, embracing IVF embryos as members. It was from this bioconstitutional moment that a new round of hESC politics started to emerge. But before turning to these issues, let us first look more closely at Law 40 and its provisions.

Italy's New "Citizen Subjects"

Law 40's main purpose was to redefine the range of admissible reproductive technologies, the organizations that would be allowed to offer such technologies, and the subjects who would be allowed to resort to them. In these efforts, the legislators took "natural boundaries" as their reference frame, and positioned IVF embryos as gatekeepers of this frame. These were now included in the community of citizen subjects whose rights and well-being were supervised by the state and its authorities (Filippini 2004).[5] The first article of the law spells out this programmatic purpose. Specifically, it states that Italians would be allowed to turn to "techniques of medically assisted procreation . . . to facilitate the solution of reproductive problems stemming from human sterility or human infertility under the conditions and following the directions made in this law which ensures the rights of all involved subjects including [the rights of] the conceived: (*concepito*)" (Repubblica Italiana 2004, Art. 1). The following articles set out a catalog of requirements and restrictions that can be read as the translation of this programmatic purpose into a list of substantive rights and sanctions. The first right of Italian embryos was the right to be created solely for the sake of embryo transfer. The law states that no more than three embryos may be produced and that all these embryos must immediately be transferred into their mother's womb. All potential technical interventions that threaten to divert embryos from their "natural" course from Petri dishes to wombs were prohibited as the law explicitly stated that embryos must not be "manipulated," "selected," or frozen for storage

(Testa 2006; Testa, chapter 4, this volume).[6] In addition, the law enshrined the embryo's right to be born into a particular family structure, consisting of one mother and one father. It prohibited the insertion of the genes of a third party through sperm and egg donation, made surrogate motherhood a punishable offense, and limited access to IVF to heterosexual couples "in a potentially fertile age" (see Marchesi 2007).

With the promulgation of the law, embryos dwelling in Italian laboratories were disentangled from their relationship with their prospective mothers and fathers and removed from the control of their physicians. They ceased to be private objects. Instead, they were embraced as public citizen subjects and put under the guardianship of the state. As such, they were effectively nationalized. In their attempt to redraw familial boundaries, the legislators ended up with a bioconstitutional innovation.

The legal birth of the new citizen subject had major implications for Italian fertility laboratories, which were no longer allowed to offer gamete donation, or preimplantation genetic diagnoses. For hESC research, this implied that Italian scientists could no longer procure stem cells from Italian embryos. Using one of Italy's new citizen subjects for research purposes became a criminal offense punishable with up to six years of imprisonment and a fine between €50,000 and €150,000 (Art. 13, Sec. 4).[7] As the law did not mention hESCs themselves, Italian scientists continued to be free to work on hESC lines imported from abroad and thus derived from "foreign" embryos, but Italian embryos were no longer at their disposal. Indeed, as the creation of surplus embryos was now illegal, such embryos were now prevented even from coming into existence. However, as I will show in the next section, the law was far from the end of the Italian politics of hESC research; in contrast, it marked its beginning. Though it restricted the number of hESCs in Italian laboratories, it provided the conditions for them to become a matter vested with collective meaning.

Rise of a "Stem Cell Collective"

Many welcomed Law 40 as the end of Italy's Wild West of reproduction (Arachi 2004), but it also met with harsh criticism. Many denounced it as a "Catholic law," with which Italy's political representatives had proved to be more attentive to the expectations of religious hierarchies than to the rights and needs of their electoral constituencies (Mafai 2004). What made the law even more disturbing was that some citizens had to pay for this religious favoritism with their bodies or indeed their lives. Infertile couples and sterile women argued that the law had seriously decreased their chances of becoming parents, leaving only wealthy Italians free to go abroad for fertility treatments that were now illegal in Italy. Carriers of genetic diseases argued that by prohibiting preimplantation genetic diagnosis the state had infringed on their right to govern their genetic destinies (ADUC 2004a). The public sphere also became populated by the damaged bodies and denied hopes of other groups who argued that the promulgation of the law amounted to a "death penalty," as a letter to the editor in the daily *Corriere della Sera* put it. This letter went on: "I do not see how you could otherwise define the destiny to which hundreds of thousands of persons suffering from terrible degenerative disease (such as Alzheimer's disease, and amyotrophic lateral sclerosis) are condemned to, [patients] whose only hope for recovery consists in scientific research with stem cells" (Pizzato 2004).

A similar language of denied hopes with deadly consequences was mobilized by the Luca Coscioni Association (*Associazione Luca Coscioni*), a collective of politicians, physicians, scientists, infertile couples, and patients suffering from genetic diseases and chronic conditions. Association president Luca Coscioni, who suffered from amyotrophic lateral sclerosis (also known as Lou Gehrig's disease), defined himself as an expert on "bioethics through [his] own skin" (Coscioni 2003, 122). His disease confined him to a wheelchair, and he was able to speak only with the help of a special computer and an electronic synthesizer. He denounced the law as having "trampled on human dignity, disregarded hope" and "sentenced thousands of patients to death." He put his "suffering body" and "metallic voice" (ADUC 2004b) into a struggle that for him was "a battle for freedom" (Coscioni 2003, 46–47). Drawing explicitly on a discourse of rights, Coscioni claimed, "Every patient has the civil right to avail himself of the progress of scientific research. The rights to health, to recovery or, anyhow, to a reduction of suffering, have to be respected. [They] must not be violated by a dogma-law of a State that blocks the freedom to conduct research, of finding the world of life and nature in the name of faith" (qtd. in Quaranta 2006).

On the association's web page, patients and their families from all over Italy described their experiences and encounters with Law 40, relating how it had "struck their lives." Speaking in the name of their physical vulnerability and mobilizing their damaged bodies, they acted as "biological citizens" in a way that was reminiscent of Adriana Petryna's biological citizens in post-Soviet Ukraine (Petryna 2002; Rose and Novas 2005). They demanded freedoms and civil rights from the state that they argued had deprived them of these rights, such as the right to make their own ethical decision on how to conduct their life, the right to cure their sick bodies, and the right to have at least a chance to invest their hopes in scientific progress (Associazione Luca Coscioni 2004). These citizens aligned themselves with researchers and scientists who left their laboratories to demonstrate that they were not the "Frankensteins" they were depicted as being (Fallaci 2005). They were supported by other adult citizens, who had not been directly affected by the law but who nevertheless agreed that it was an unjust or even a "cruel" act (Mafai 2004).

This "stem cell collective" solidified its protest by making use of a device within the Italian constitution that enables a law to be repudiated by a referendum. In summer 2004, the Luca Coscioni Association and other groups collected enough signatures for five referendum petitions. In January 2005, the Constitutional Court approved four of them. Three sought to relax the tight restrictions on IVF practices in Italy.[8] The fourth sought to "free research and the hope for cures of millions" (Associazione Luca Coscioni 2005) by abrogating the law's restrictions on hESC research. With the court's approval of the referendum hESCs ceased to be a "nonissue." Instead, they were now firmly positioned on Italy's political stage and became the trigger and signifier of a battle for freedom against an "oppressive" state that had allegedly sacrificed the rights and hopes of its adult citizens on the altar of its Catholic ally.

Yet in the five months between the approval of the referendum in January 2005 and the polling days in June 2005, the focus of attention shifted from the denied rights, freedoms, and hopes of adult citizens to the identity and rights of Italian embryos. The personal narratives and embodied experiences of citizens whose lives had been hampered by the enactment of the law were silenced and replaced with the authority of scientific facts and the pronouncements of philosophers and bioethicists. Clergymen were not absent from these debates, yet God and religious truths were surprisingly silent. Italy's national soul searching on the pitfalls of stem cell technologies took the form of lessons about human genomes and excursions into human biology.

Human Biology "Cannot Be Put to a Vote"

Between January 2005, when the Constitutional Court allowed the referendum to go forward, and the polling days on June 12–13, a broad coalition of actors defended the law, because it "has finally put an end to the so called 'Wild West of procreation,' giving every child the guarantee of a human life and the protection of a true family" (Comitato Scienza & Vita 2005).

They reframed the issues at stake. A series of videotapes entitled "It's Life," distributed across Italy by the newly established Science & Life Association (*Associazione Scienza & Vita*), stated that the key question the Italian electorate should reflect upon was

the embryo, on who it is, on its rights, and on whether it should be regarded as a ball of cells with which one can experiment, or which can be manipulated or even trashed, when it is useless or when one presumes that it might be sick or imperfect. This is the question that is put to us, now that we are pressured by queries that would want us to go to the polls pretending the right to decide when life is really life and from when it has a value. (SATduemila and Scienza & Vita 2005b)

Although the law's defenders wanted Italian voters to consider what the embryo is, the answer to this ontological question, it appeared, was neither open to dispute nor to be solved by giving adult citizens their say. This was a matter for science, and the answer was clear. As Paola Binetti, one of the presidents of the Science & Life Association explained, "in this case biology" was "incontrovertible" (SATduemila and Scienza & Vita 2005a). As the result of the combination of the genes of its father and its mother, the embryo's genome provided all the necessary information to make an embryo become a full-fledged and unique human being. It was the embryo's genome that rendered it "one of us" (*uno di noi*). Further, the status of the embryo as "one of us" was also derived from the fact that we "have all been embryos" (Della Fratina 2005). "We have all been embryos" became one of the campaign slogans of the defenders of Law 40. As the parliamentarian Francesco Paolo Lucchese had already explained in February 2004, "One cannot freeze one of us. The embryo cannot be manipulated for research purposes, because this would be as if we put one of us at the disposition of research" (Camera dei Deputati, 2004, 31).

Nor could the life of one of us be put to a vote. "Life cannot be put to a vote" turned out to be the key slogan of the campaign. It had a double meaning. On the one hand, it advised the Italian electorate to recognize the intrinsic worth and value of human life and to dismiss not only the content of the referendum but the process itself. On the other hand, it had a very pragmatic significance. Following Cardinal Ruini's strategy, the defenders of the law based theirs not merely on the electorate's self-identification with Italian embryos but also on the known reluctance of many Italians to cast their votes in such polls. By claiming that "life cannot be put to a vote," they tried to maximize their output, capitalizing not only on Italians' understanding of genomes and rights, but also on past experiences with the political inertia of some of their fellow citizens.

Moreover, they argued that the stem cell collective was tragically misled and also misinforming the electorate with the argument that Italy's laws infringed on their rights and lives. Scientist Angelo Vescovi explained on one of the posters that were widely displayed throughout Italy that it was neither "licit" nor "helpful" to "destroy embryos to cure diseases." In a further bow to scientific authority, the campaign argued that "results have not yet been achieved with embryonic stem cells. In contrast, excellent results have been achieved with adult stem cells." Drawing a line between scientifically proven and morally sound adult stem cells, and scientifically uncertain and certainly immoral embryonic ones, Vescovi declared that he would not vote, inviting his lay fellow citizens to follow his scientifically grounded example (Minerva 2004; Vescovi 2005).

On the two voting days, the strategy of abstention proved successful— only 25.7 percent of the Italian electorate went to the polls. It was thus a mere historical footnote that 88.0 percent of those who did cast a vote wished to remove the restrictions on hESC research.[9] The threat posed to the Italian embryo regime was averted. No embryo would die for Italian hESC research; each was "one of us."

However, as I will argue in the next section, the results of the referenda were not merely a defeat for the rights of those Italians who invested their hopes in hESC research, or just a victory for the ontological inscription of embryos as "one of us." Rather, both entities worked as metaphors of a larger dispute. What seems to be a small record in the history of the Italian Republic was another feature of this bioconstitutional moment that was not staged in courtrooms or in legislative bodies but dispersed among newspaper articles, websites, and voting booths. This was constitutionalism with a small "c" (Jasanoff, chapter 1, this volume) but Politics with a capital "P," as it opened up the very foundational categories of Italian (bio) politics, focusing on issues such as what it means to be a citizen, whose rights, and of what sort, the state should protect, and what life is and what it should be. Sharply contrasting understandings of these issues emerged.

Life at Issue

The view of human life held by the stem cell collective was most effectively embodied by Luca Coscioni. His biography was often told in the following way: "He was training to participate in the New York Marathon, when he discovered he suffered from a disease that is today not curable and that paralyzes him, step by step, and renders him incapable of speaking" (Quaranta 2006).

Coscioni's life had played a cruel trick on him, depriving him of the ability to walk and to speak. Technology gave him back the voice that the lottery of life had stripped from him, and hESC research gave him hope of survival. For the stem cell collective, human life left to itself was far from perfect: it was in need of interventions, enhancements, and improvements. Life was not a destiny that had to be accepted, but something that could and should be acted upon by citizens and scientists, for the sake of repairing ailing bodies and improving life itself. The stem cell collective referred to a "private" life that belonged to the individuals who embodied that life, who had the right to speak for it, and to decide and act on it. In their vision, the state as a "good shepherd" had the duty to protect even these defective citizens and guarantee them the freedom and right to normalize their bodies and enhance their diminished lives.

The vision of human biology held by the defenders of the law was very different. Symbolically embodied by human embryos, their notion of life could not speak on its own and was therefore in need of somebody who would speak for it and try to read it, without attempting to rewrite it. This life seemed to be God-given, yet in the absence of a universally shared faith in transcendental authority, it was argued to be inherent in a more worldly authority—the human genome. Nobody could "pretend" to have the right to decide on the worth or value of this life or to put it "to a vote" in the daily actions or inactions of patients, couples, physicians, and scientists or in a national referendum. To the contrary, the state and its institutions had the duty to protect this vulnerable life from infringement by maverick doctors and scientists, and from egoistic patients and citizens. This understanding of "deprivatized," "naturalized," and "nationalized" life resembled the vision of human biology enshrined in Law 40. This was the vision that the "stem cell collective" wanted to contest, and it was this vision that survived the referendum.

How did this vision arise in the first place? Why did the Italian Parliament end up nationalizing IVF embryos even before the referendum gave them new birth as "one of us"? And what was the role of the Catholic Church in promoting that result? In a concluding section, I will briefly explore these questions, going back to the controversies on reproductive medicine that led to the enactment of Law 40.

Italy's "Wild West" of Reproduction

Conflicts about embryos and collective puzzling about what they are and what they should be are not an Italian prerogative. These tiny entities became the stuff of big politics in many countries at the turn of the twenty-first century. In the United States, former President George W. Bush earned intense criticism for his fervent opposition to federal funding for most hESC research (Jasanoff 2005b; Gottweis and Prainsack 2006). In Europe, Germany was at pains to reconcile the promises of hESC research with its constitutional commitments to the protection of human life and human dignity (Gottweis 2002; Jasanoff 2005b; Herrmann 2009). The birth of the Italian embryo and its regulatory regime shares some points with these other stories, but also has some distinguishing features. Although the effect of the Italian legal stipulations are almost congruent with the provisions in Germany's Embryo Protection Act (*Embryonenschutzgesetz*) of 1990—both states in effect enacted an "ontological prohibition" to surplus embryos, "keeping entities potentially disruptive of the moral order from ever coming into being" (Jasanoff 2005b, 146)—the way in which they arrived at such a prohibition differed markedly. In Italy, pre-existing constitutional commitments did not drive the establishment of the Italian embryo as citizen subjects. Rather, appeals to constitutional rights at first threatened this very project and later helped to undermine it.[10] In turn, religious foundations were certainly more salient in Roman Catholic Italy than in confessionally more heterogeneous Germany (Hanafin 2006, 2007; Schiffino et al. 2009). When the Italian Parliament enacted Law 40, its members followed their personal conscience, which meant that they voted on religious rather than party lines. Catholic moral teaching also lies barely concealed under the surface of the law's stipulations. Yet, intriguingly, these teachings remained tacit and not openly discussed. In contrast, legislators poured great effort into explaining that they appealed not to a divine truth but to a secularized version of this authority. A seemingly less suspicious nature was the "storyline" (Hajer 1995) that bridged a transcendental Catholic morality and a worldly Italian sense of right and wrong. It also connected legislators' search for ordering lines with the boundaries that were already deployed in judgments of right and wrong in newspapers, television shows, bioethics committees, and courtrooms (Bucchi 1999; Neresini 2000; Testa 2006; Testa, chapter 4, this volume). Though the frequency of the appeals to nature both inside and outside the parliament may itself have Catholic roots, the legislative position gained its power by concealing these contingent roots, as well as through the authority of the scientific and bioethics spokespersons who were mobilized to convey nature's truths.

Members of the Italian Parliament were perhaps most influenced by previous debates about assisted reproduction. Since the early 1980s, reproductive technologies had been widely deployed in Italian laboratories and clinics (Bonaccorso 2004; Valentini 2004), but they were nevertheless regarded with suspicion. The new technologies, it seemed, not only fixed medical problems and restored the natural reproductive order, but also created new ways to procreate, enabling the birth of children with multiple mothers, changing the age of motherhood, engendering (un)familiar tragedies, and undermining the natural boundaries of the family (Marchesi 2007; Neresini 2000). Despite public unease, policy makers at first preferred to coexist with these unregulated "monsters" (Jasanoff 2005b, 151), rather than to venture into finding a common basis for taming them, thereby officially taking note of their existence and risking a clash with the Catholic hierarchy. The new ambiguous realities were framed as a problem that was not solvable through party politics or governable through acts by partisan governments. Instead, a national law, to be deliberated on the basis of the souls and conscience of the representatives of the Italian nation, was deemed the only appropriate instrument of governance.

As legislative momentum grew, commentaries flourished about the ambiguous entities that were emerging from Italy's private fertility laboratories. These set the stage for the subsequent ordering attempts by Parliament. They depicted assisted reproduction as a sphere without rules and norms, threatening the natural order. For example, as a supermarket of "homosexuals who fertilize lesbians, multiple births at request and, virgins who birth children" (L'Espresso, 1995, qtd. in Valentini, 2004, 98). In the term increasingly deployed from the early 1990s, this was Italy's "Wild West" of reproduction (Valentini 2004; Hanafin 2006, 336; Marchesi 2007).

This discursive space was the medium in which Italy's embryonic citizens could grow and live. They were framed as the embodiment of a vulnerable nature that was under attack, and as—lacking their own voice—in need of the law's protective intervention. Embryos were construed as little children who were being mistreated by maverick medical professionals or abandoned by egoistic parents, and hence as the primary victims of the wild reproductive frontier. In turn, they now became the reference point in legislators' attempts to restore order. The protection of the embryos' rights became the guiding rationality that inspired the legislation and the primary argumentative device that structured legislative deliberation. Their right to be produced only for the sake of embryo transfer, not to be disturbed on this path, and their right to be born into a family with "natural" parents were rooted in genetic ties. Sound biological boundaries, it seemed, were needed not only to shelter the identity and growth of the embryo but also to regenerate the entire nation.

Law 40 did not interfere with Italy's Abortion Act, which permits women to end an unwanted pregnancy in public clinics supervised by medical professionals (Calloni 2001; Galeotti 2003; Hanafin 2006; Hanafin 2007). Embryos and fetuses in their prospective mothers' wombs, and hence in their "natural space," remained outside the legislators' gaze. Drawing a discursive line between what happened in those natural spaces and what was artificially made in laboratories, legislators framed the lab-created embryos as "national subjects" in need of state protection. These now belonged to the moral community of Italian citizens, and the state and its institutions would take care of their safe development, precisely by limiting their proliferation. All actions by humans that threatened the embryos' safe and undisturbed trajectory to their natural place—that is, their prospective mothers' wombs—were banned. Natural limits that had become blurred by technology were reconfigured as legal boundaries. Italy's nationalized embryos were subject only to "nature's own will," now reinforced through legal boundaries. In short, a new citizen subject and a new reproductive order were coproduced, mutually reinforcing one another.

Conclusions

In this article, I have used Italian hESC politics as a window to explore a wider reshuffling of Italy's biopolitics from which hESC policies emerged. I started with a discussion of an early period of hESC politics, in which hESCs were largely a nonissue. This changed when the Italian Parliament passed Law 40, which endorsed early human IVF embryos as citizen subjects whose rights the state was determined to protect. Therewith, hESCs became the symbols of a battle for rights that peaked in an abrogative referendum. As we have seen, the micro politics of the referendum seeking to relax this law recapitulated Italy's macro hESC politics. Though Parliament, in passing the law, transformed embryos into citizen subjects, the referendum redefined them as "one of us." In both instances, collective sense making on what the embryo *is* and what it *should* be provided a means to tame unruly conflicts. Soul searching merged into genome searching, as ontologies and moral orders were coproduced. But, the referendum was also a micro performance of Italy's hESC politics in another sense. The attempted liberation of Italian hESC research was not defeated by active disapproval but by a silent majority that refrained from asserting its voice, just as hESCs had once been neither endorsed nor banned but silently excluded from the communities cared for by Italian authorities.

This story encourages us to refrain from reducing contemporary biopolitics to a single, coherent logic. Collectives such as states have reacted differently to the new possibilities of an age in which human biology is no longer seen as an a priori script that precedes human actions and interventions, but as something that is plastic and open to human intervention (Jasanoff 2005a; Franklin 2007; Landecker 2007; Rose 2007). Some have placed increasing hope on the self-governance of knowledgeable and responsible citizens in the face of new developments in biology and biotechnology (Rose 2001; Gottweis et al. 2004; Lemke 2004). But the politics of hESC research in Italy paints a rather different picture. Instead of seizing the opportunities for reinventing adult citizens and their identities and responsibilities, Italian legislators preferred to retreat to legally safe natural facts, and to restore order by referring to the naturalness of genes and genomes. Old and accepted facts of life were reasserted and made undeniable by the helping hand of the law. This did not convince all Italians, but the reaffirmation of a divine order in secular and scientific terms certainly satisfied the Catholic Church as much as it pleased Cardinal Camillo Ruini.

Notes

1. This chapter is adapted from an article that appeared as Metzler 2007 (reproduced with permission of Palgrave Macmillan). It draws on research conducted in the context of the research project PAGANINI (Participatory Governance and Institutional Innovation), funded under the Sixth EU Framework Programme, Contract no. CIT2-CT-2004-505791 (<http://www.univie.ac.at/LSG/paganini/>). I would like to thank Herbert Gottweis for his continuous support and Sheila Jasanoff for her perceptive comments on draft versions of this article and for her careful editorial guidance.

2. I base my interpretation on a study in which I traced hESCs as they appeared on Italy's political stage and were imbued with collective meaning (Fischer 2003; Gottweis 1998). I did not make an a priori preference of a particular kind of data. I drew on media data from newspapers, television shows and radio broadcasts, on policy documents, and on semistructured interviews that I conducted in Italy with policymakers, bioethicists, scientists, and patient activists in June 2005 and September 2006.

3. Although the ban on human cloning was extended up to the promulgation of Law 40 in February 2004, the ban on animal cloning was subsequently first relaxed and then altogether dropped in December 2001 (Ministero della Salute 2001).

4. As explained in more detail by Testa (chapter 4, this volume), the "personalization" of hESCs consists of the combination of hESC technologies with "somatic cell nuclear transfer" (SCNT). SCNT involves the transfer of the nuclear genome of a somatic cell, such as an udder cell or a skin cell, into an enucleated oocyte or egg cell, which is then stimulated to start to divide and develop like a "standard" embryo. The resulting embryo shares the nuclear genome with the donor of the

somatic cell. hESCs derived from such a "cloned" embryo would therefore be genetically perfectly compatible with the donor of the somatic cell (Hochedlinger and Jaenisch 2006; Wilmut and Highfield 2006).

5. For a discussion on the effects of the politics of "embryo citizenship" on women's rights, see Hanafin 2006 and Fenton 2006.

6. An exclusion were made for those cases of "major force" in which unforeseen health problems of the woman did not allow an immediate embryo transfer. In those cases, embryos could be frozen, yet still had to be transferred as soon as possible. In the application of the law, syndromes induced by the overstimulation of oocytes proved to be one of the causes in which medical professionals refrained from transferring embryos, and decided to freeze them instead (Ministero del Lavoro, della Salute e delle Politiche Sociali 2009).

7. Producing embryos through "somatic cell nuclear transfer" was transformed from a breach of an ordinance into a criminal offense that was punishable with up to twenty years of imprisonment, a fine between €600,000 and €1,000,000, and a lifelong exclusion from practicing the profession (Art. 12, Sec. 7).

8. The second referendum "for the protection of the health of the woman" sought to cancel the ban on embryo freezing and all provisions that outlaw preimplantation genetic diagnosis. The third sought to reaffirm "[women's] auto determination and the protection of women's health." The fourth and final referendum sought to reintroduce gamete donation to the range of permissible practices in Italian fertility laboratories (Associazione Luca Coscioni 2005; Comitato Scienza & Vita 2006).

9. For the detailed results of the consultations, see the booklet published by the Science & Life Association (Comitato Scienza & Vita 2006).

10. In general, Italian Constitutional Law and its interpretation through the Italian Constitutional Court proved to be the single most important institutional ally of Italian women and the protection of their reproductive rights (Galeotti 2003). Debates that led to Italy's Abortion Act in 1978 were kicked off by a decision of the Constitutional Court that declared the norms governing abortions in Italy unconstitutional. Enacted during the period of Italian Fascism, these reflected prevalent eugenic understandings. They criminalized abortion as one of a series of the "Crimes against the Integrity and Health of the Stock" (Hanafin 2009, 228–229). Though the Constitutional Court argued that protection of the "conceived"—now for the sake of the life of the "conceived" and not for the sake of the vitality of the "stock"—rested on constitutional foundations, it stressed that this aim had to be balanced with other protected "goods." In particular, the Court argued that "there is no equivalence between the right not only to live but also to health owned by who is already person, as the mother, and the protection of the embryo who has yet to become a person" (Corte Costituzionale 1975; Hanafin 2009, 231; Galeotti 2003, 116). More than thirty years later, the constitutionally protected right to health again led the Constitutional Court to declare parts of Law 40 unconstitutional (Corte Costituzionale 2009). In its decision of April 2009, the Court declared unconstitutional those provisions of the law declaring that no more than three embryos may be produced, that all produced embryos had to be transferred to the prospective mother's wombs, and that none might be frozen.

After this judgment, Law 40 continues to stipulate that "no more than the necessary number of embryos" may be produced. However, what "necessary" means is no longer governed by law but by medical judgment.

References

Abbott, Alison. 2009. Italians sue over stem cells. *Nature* 460:19.

ADUC. 2004a. Italia. Cartello per dire no alla legge sulla Procreazione Medicalmente Assistita. *Notiziario Cellule Staminali* III (57). Retrieved from <http://salute .aduc.it/staminali/notizia/italia+cartello+dire+no+alla+legge+sulla_57855.php>

ADUC. 2004b. Italia. Luca Coscioni: aboliamo la legge che vieta la ricerca, insieme. *Notiziario Cellule Staminali* III (57). Retrieved from <http://staminali.aduc .it/php_newsshow_0_3059.html>.

Arachi, Alessandra. 2004. Fecondazione, il Parlamento vara le regole. *Corriere della Sera* (February 20): 3.

Associazione Luca Coscioni. 2005. *Io voto 4 volte sí*. Roma: Associazione Luca Coscioni.

Bonaccorso, Monica. 2004. Making connections: Family and relatedness in clinics of assisted conception in Italy. *Modern Italy* 9 (1): 59–68.

Bucchi, Massimiano. 1999. *Vino, alghe e mucche pazze. La rappresentazione televisiva delle situazioni a rischio*. Roma: Radiotelevisione Italiana.

Calloni, Marina. 2001. Debates and controversies on abortion in Italy. In *Abortion politics, women's movements, and the democratic state. A comparative study of state feminism*, ed. Dorothy McBride Stetson, 181–203. Oxford: Oxford University Press.

Camera dei Deputati. 2004. Resoconto stenografico dell'Assemblea, Seduta n. 421 del 10 Febbraio 2004. Retrieved from <http://www.camera.it/_dati/leg14/lavori /stenografici/sed421/s000r.htm>

Cattaneo, Elena, Elisabetta Cerbai and Silvia Garagna. 2009. La figuraccia di chi odia le staminali. *La Stampa* (July 8): 23–24.

Cazzullo, Aldo. 2003. Januzzi lo sconfitto: è libertà di coscienza ma la chiamano Far West. *Corriere della Sera* (December 12): 6.

Comitato Scienza & Vita. 2005. Comitato "Scienza & Vita" per la Legge 40/2004. Un doppio no alla menzogna (leaflet).

Comitato Scienza & Vita. 2006. *Referendum 2005 sulla Fecondazione Medicalmente Assistita. Essere umani dall'inizio alla fine. Quattro mesi vissuti intensamente per affermare il primato della vita*. Pomezia: La Fenice Grafica.

Corte Costituzionale. 1975. N. 27, Sentenza 18 Febbraio 1975. *Gazzetta Ufficiale* 55 (February 26). Retrieved from <http://www.cortecostituzionale.it/actionScheda Pronuncia.do?anno=1975&numero=27>.

Corte Costituzionale. 2009. Sentenza N. 151, Anno 2009. *Gazzetta Ufficiale, 1ª Serie Speciale Corte Constitutionale* 19 (May 13). Retrieved from <http://www .cortecostituzionale.it/actionSchedaPronuncia.do?anno=2009&numero=151>.

Coscioni, Luca. 2003. *Il maratoneta. Dal caso pietoso a caso pericoloso. Storia di una battaglia di libertà*. Viterbo: Stampa alternativa.

De Bac, Margherita. 2003. Fecondazione artificiale. Arriva al Senato la legge più severa d'Europa. *Corriere della Sera* (July 10): 18.

Della Fratina, Giannino. 2005. Formigoni: "Mi astengo. Siamo tutti ex embrioni." *Il Giornale* (June 8): 6.

Fallaci, Oriana. 2005. Noi cannibali e i figli di medea. *Corriere della Sera* (June 3): 1, 8–9.

Fenton, Rachel Anne. 2006. Catholic doctrine versus women's rights: The new Italian law on assisted reproduction. *Medical Law Review* 24:73–107.

Filippini, Nadia Maria. 2004. Il corpo dominato e la personificazione dell'embrione: una prospettiva storica. In *Un'appropriazione indebita. L'uso del corpo della donna nella nuova legge sulla procreazione medicalmente assistita*, ed. AA.VV., 97–112. Milano: Baldini Castoldi Dalai.

Fischer, Frank. 2003. *Reframing public policy. Discursive politics and deliberative practices*. Oxford: Oxford University Press.

Franklin, Sarah. 2007. *Dolly mixtures: The remaking of genealogy*. Durham: Duke University Press.

Galeotti, Giulia. 2003. *Storia dell'aborto*. Bologna: Il Mulino.

Gottweis, Herbert. 1998. *Governing molecules. The discursive politics of genetic engineering in Europe and the United States*. Cambridge, Mass.: MIT Press.

Gottweis, Herbert. 2002. Stem cell policies in the United States and in Germany: Between bioethics and regulation. *Policy Studies Journal: the Journal of the Policy Studies Organization* 30 (4): 444–469.

Gottweis, Herbert, Wolfgang Hable, Barbara Prainsack, and Doris Wydra. 2004. *Verwaltete Körper. Strategien der Gesundheitspolitik im internationalen Vergleich*. Wien, Köln and Weimar: Böhlau Verlag.

Gottweis, Herbert, and Barbara Prainsack. 2006. Emotion in political discourse: Contrasting approaches to stem cell governance—The U.S., UK, Israel, and Germany. *Regenerative Medicine* 1 (6): 823–829.

Gottweis, Herbert, Brian Salter, and Chatherine Waldby. 2009. *The global politics of embryonic stem cell science*. London: Palgrave MacMillan.

Hajer, Maarten A. 1995. *The Politics of environmental discourse. Ecological modernization and the policy process*. Oxford: Oxford University Press.

Hanafin, Patrick. 2006. Gender, citizenship, and human reproduction in contemporary Italy. *Feminist Legal Studies* 14 (3): 329–352.

Hanafin, Patrick. 2007. *Conceiving life: Reproductive politics and the law in contemporary Italy*. Aldershot, UK: Ashgate.

Hanafin, Patrick. 2009. Refusing disembodiment: Abortion and the paradox of reproductive rights in contemporary Italy. *Feminist Theory* 10 (2): 227–244.

Herrmann, Svea Luise. 2009. *Policy debates on reprogenetics: The problematization of new research in Great Britain and Germany*. Frankfurt, N.Y.: Campus Verlag.

Hochedlinger, Konrad and Rudolf Jaenisch. 2006. Nuclear reprogramming and pluripotency. *Nature* 441 (June 29): 1061–1067.

Jasanoff, Sheila. 2005a. *Designs on nature: Science and democracy in Europe and the United States*. Princeton: Princeton University Press.

Jasanoff, Sheila. 2005b. In the democracies of DNA: ontological uncertainty and political order in three states. *New Genetics & Society* 34 (2): 139–155.

La Repubblica.it. 2005. "Ruini festeggia la vittoria, 'Italiani, popolo maturo.'" *La Repubblica.it* (June 13). Retrieved from <http://www.repubblica.it/2005/f/sezioni /politica/dossifeconda6/rearui/rearui.html>.

Landecker, Hannah. 2007. *Culturing life: How cells became technologies*. Cambridge, Mass.: Harvard University Press.

Lemke, Thomas. 2004. *Veranlagung und Verantwortung. Genetische Diagnostik zwischen Selbstbestimmung und Schicksal*. Bielefeld: Transcript Verlag.

Mafai, Miriam. 2004. Una norma che ignora il paese reale. *La Repubblica* (February 11): 1.

Marchesi, Milena. 2007. "From adulterous gametes to heterologous nation: Tracing the boundaries of reproduction in Italy." *Reconstruction* 7 (1). Retrieved from <http://reconstruction.eserver.org/071/marchesi.shtml>.

Metzler, Ingrid. 2007. "Nationalizing embryos": The politics of human embryonic stem cell research in Italy. *BioSocieties* 2 (4): 413–427.

Minerva, Daniela. 2004. Da laico vi dico: è una barbaria. *L'Espresso* (August 24).

Ministero del Lavoro, della Salute e delle Politiche Sociali. 2009. Relazione del Ministro del Lavoro, della Salute e delle Politiche Sociali al Parlamento sullo Stato di Attuazione della Legge Contenente Norme in Materia di Procreazione Medicalmente Assistita (Legge 19 Febbraio 2004, N. 40, Articolo 15). Retrieved from <http://www.ministerosalute.it/imgs/C_17_pubblicazioni_944_allegato.pdf>.

Ministero della Sanità. 2000. Relazione della Commissione di studio sull'utilizzo di cellule staminali per finalità terapeutiche. Retrieved from <http://www.salute .gov.it/imgs/C_17_bacheca_10_listaelencodocumenti_elenco1_listadocumenti _documento0_listafile_file0_linkfile.pdf>.

Ministero della Salute. 2001. Ordinanza 21 dicembre 2001. Proroga dell'efficacia dell'ordinanza concernente il divieto di pratiche di clonazione umana. *Gazzetta Ufficiale* 30 (February 5, 2002): 36.

Ministro della Sanità. 1997. Ordinanza 5 marzo 1997. *Gazzetta Ufficiale* 55 (March 7).

Neresini, Federico. 2000. And man descended from the sheep: the public debate on cloning in the Italian press. *Public Understanding of Science* 9:359–382.

Petryna, Adriana. 2002. *Life exposed: Biological citizens after Chernobyl*. Princeton: Princeton University Press.

Pizzato, Paolo. 2004. Fecondazione assistita come la pena di morte. *Corriere della Sera* (February 1). Retrieved from <http://sitesearch.corriere.it/engineDocument Servlet.jsp?docUrl=/documenti_globnet5/mondo_corriere/Italians/2004/02/12 /03040212.xml&templateUrl=/motoriverticali/italians/risultato.jsp>.

Politi, Marco. 2005. Ruini, il giorno del trionfo. Ho fatto solo il mio dovere. *La Repubblica* (June 14): 2.

Quaranta, Pasquale. 2006. "TEMPO SCADUTO! Le parole di Luca Coscioni." *il "Cassero" magazine* (March 9). Retrieved from <http://www.lucacoscioni.it /tempo-scaduto-le-parole-di-luca-coscioni>.

Ramjoué, Celina, and Ulrich Klöti. 2003. Assisted reproductive technology policy in Italy: Explaining the lack of comprehensive regulation. In *Comparative Biomedical Policy: Governing Assisted Reproductive Technologies*, ed. Ivar Bleiklie, Malcolm L. Goggin, and Christine Rothmayr, 42–63. London: Routledge.

Repubblica Italiana. 2004. Legge 19 febbraio 2004, n. 40, Norme in materia di procreazione assistita. *Gazzetta Ufficiale* 45 (February 24). Retrieved from <http:// www.camera.it/parlam/leggi/04040l.htm>.

Rose, Nikolas. 2001. The politics of life itself. *Theory, Culture & Society* 18 (6): 1–30.

Rose, Nikolas. 2007. *The politics of life itself. Biomedicine, power, and subjectivity in the twenty-first century.* Princeton: Princeton University Press.

Rose, Nikolas, and Carlos Novas. 2005. Biological citizenship. In *Global Assemblages. Technology, Politics and Ethics as Anthropological Problems*, ed. Aihwa Ong and Stephen J. Collier, 439–463. Malden, Mass.: Blackwell Publishing.

SATduemila and Scienza & Vita. 2005a. *Dalla diagnosi prenatale alla diagnosi preimpianto.* Roma: Rete Blue S.p.A. (videotape).

SATduemila and Scienza & Vita. 2005b. *Sperimentazione sull'embrione: diritti del concepito e salute della donna.* Roma: Rete Blue S.p.A. (videotape).

Schiffino, Nathalie, Celina Ramjoué, and Frédéric Varone. 2009. Biomedical policies in Belgium and Italy: From regulatory reluctance to policy changes. *West European Politics* 32 (3): 559–585.

Testa, Giuseppe. 2006. Che cos'è un clone? Pratiche e significato delle biotecnologie rosse in un mondo globale. In *Cellule e cittadini. Biotecnologie nello spazio pubblico*, ed. Massimiano Bucchi and Federico Neresini, 141–162. Milano: Sironi Editore.

Thomson, James A., Joseph Itskovitz-Eldor, Sander S. Shapiro, Michelle A. Waknitz, Jennifer J. Swiergiel, Vivienne S. Marshall, and Jeffrey M. Jones. 1998. Embryonic stem cell lines derived from human blastocysts. *Science* 282 (5391): 1145–1147.

Valentini, Chiara. 2004. *La fecondazione proibita.* Milano: Giangiacomo Feltrinelli Editore.

Vescovi, Angelo. 2005. Bugie staminali. *Il Foglio* (January 22): 1.

Wilmut, Ian, and Roger Highfield. 2006. *After Dolly: The uses and misuses of human cloning.* London: Little, Brown Book Group.

6

Certainty vs. Finality: Constitutional Rights to Postconviction DNA Testing

Jay D. Aronson

It is better that ten guilty persons escape, than that one innocent suffer.
—William Blackstone (Blackstone 1765–1769, 352)

No one, not criminal defendants, not the judicial system, not society as a whole is benefited by a judgment providing a man shall tentatively go to jail today, but tomorrow and every day thereafter his continued incarceration shall be subject to fresh litigation.
—Justice John Marshall Harlan II (*Mackey v. United States* 1971, 691)

Introduction

At least in theory, the American criminal justice system is designed to ensure that innocent men and women are not wrongfully convicted for crimes that they did not commit. Constitutional and procedural safeguards abound. American citizens enjoy the right to a jury trial, the right to remain silent upon questioning by the state, the right to legal counsel, the right to examine all of the state's evidence before trial, the right to cross-examine opposing witnesses, as well as an overarching right to due process. Convicted prisoners also have the right to challenge a conviction if any constitutional rights were denied during trial, and also to seek clemency from the executive authority of the jurisdiction in which they were convicted.

Despite these safeguards, defense lawyers and civil liberties advocates have been arguing for years that the American legal system is in fact fundamentally unfair and unjust. Because of power and resource imbalances, federal and state prosecutors win convictions against individuals who did not commit the crimes for which they were on trial. As a result, thousands of actually innocent people may be languishing in prisons and death rows around the country (Bedau and Radelet 1987; Borchard 1932; Gross et al.

2005; Radelet, Bedau, and Putnam 1992; Radin 1964; Scheck, Neufeld, and Dwyer 2000).

In the past, such claims were difficult to prove, primarily because of the degradation of evidence, both physical and eyewitness, and the fundamental belief in the correctness of legal decision making (Bedau and Radelet 1987; Berger 2004). However, forensic DNA analysis is increasingly being used in postconviction litigation to prove that innocent people have been wrongfully incarcerated (Scheck, Neufeld, and Dwyer 2000). More than a decade and more than 250 exonerations later,[1] the Innocence Project at the Cardozo School of Law in New York City and its sister organizations have created a moment in which long-held assumptions about the fairness and efficacy of our criminal justice system are being called into question (Aronson and Cole 2009; Berger 2004).

Still, the decisions of our criminal courts are considered to be final unless a defendant's constitutional rights were violated at trial. In a landmark 1993 case, *Herrera v. Collins*, the Supreme Court ruled that even the "actual innocence" of a prisoner (i.e., the *fact* that the person did not commit the crime for which he was convicted) was not sufficient to necessitate the reversal of a conviction. Rather, it could only serve as the "gateway though which a habeas petitioner must pass to have his otherwise barred constitutional claim considered on the merits" (*Herrera* 1993, 404). In other words, the *Herrera* majority found that the weak but widely distributed right of all Americans to legal finality and repose (the notion expressed by Justice Harlan in *Mackey*) outweighs a defendant's narrowly distributed, individual right to absolute certainty in legal decisions (the notion expressed above by Blackstone), as long as no constitutional violations led to the conviction.

Herrera raised significant legal challenges for defense lawyers hoping to use DNA test results to vacate the convictions of their clients. In many states, defense lawyers gained postconviction access to biological evidence through legislation, ad hoc agreements with prosecutors, and other legal processes. However, a major complaint made by the community seeking to overturn wrongful convictions is that there is no fail-safe right to DNA testing throughout the country. Though forty-eight states and the federal government have statutes mandating access to biological materials for postconviction DNA testing when conditions of varying stringency are met, as of July 2010 access in Oklahoma and Massachusetts still depended completely on the beneficence of government officials or the case-by-case decisions of individual judges.

According to the Innocence Project, although access to postconviction DNA testing has improved dramatically over the past decade, there are still numerous flaws and holes in coverage—even in those states that have passed statutes. Some statutes, for instance, set very high evidentiary hurdles before access is granted; others prohibit access for people who plead guilty to a crime (even though the problem of false confessions is well documented); several states do not allow defendants to appeal denials of postconviction testing; and many states do not require courts to act quickly on a request for postconviction DNA testing once it has been filed (Innocence Project 2010). In other words, there is no ironclad guarantee that any convicted person in any prison in the country could gain access to postconviction DNA testing that could prove his or her innocence. Ensuring this unfettered access is the Innocence Project's ultimate goal.

Consequently, the Innocence Project and other organizations have called for the creation of a fundamental constitutional right to postconviction testing, thus overriding the balancing and utility tests that prosecutors and courts ordinarily use to deny access to biological evidence in the name of finality and social stability. This demand is based on the claim that DNA evidence has the power to provide "cast iron scientific proof," whereas our system convicts and sentences innocent people on a regular basis based on flawed forensic evidence and unreliable eyewitness testimony (Leahy 2001). As Barry Scheck, Peter Neufeld, and Paul Dwyer wrote in their book *Actual Innocence: Five Days to Execution, and Other Dispatches from the Wrongly Convicted*, "In what seems like a flash, DNA tests performed during the last decade of the [20th] century . . . have exposed a system of law that has been far too complacent about its fairness and accuracy" (2000, xv).

Such claims, as I will show, depend crucially on concurrent acts of construction and purification[2] of scientific techniques and the knowledge they produce. In order to be elevated to the status of a constitutional right, or as the clincher of a foolproof death penalty, DNA typing must also be elevated into the ultimate identification evidence, and all others must be simultaneously downgraded (Aronson 2007; Lynch et al. 2008). This work of constructing DNA's invincibility must then be rendered invisible, so that DNA evidence can speak with the disembodied power and authority of objective truth. Put differently, we see a two-pronged story unfolding: on the one hand, DNA evidence must be made foolproof and to speak for itself; on the other hand, the constitutional right to testing must be made to seem naturally flowing from the authority of DNA evidence. These two constructions are dependent upon and intimately linked to one another.

Although the U.S. Supreme Court narrowly declined to recognize the existence of this right in its 5–4 decision in *District Attorney's Office v. Osborne* (2009), the legal arguments surrounding the case raised fundamental questions about the reframing of rights through technological change. Ultimately, though further establishing DNA evidence as the "gold standard" of proof whose validity and accuracy are superior to all other forms of forensic evidence, DNA testing failed to dislodge process as the ultimate legitimator of finality in the courts. Although some judges (including four Supreme Court justices) were eager to modify existing legal procedures based on the authority of DNA evidence, others sought to defend the sanctity of process in law from incursions by alternative, extralegal sources. At stake was the means by which our legal system can best balance the desire to provide justice to individual defendants and the need to maintain social order: through novel technological practices or well-entrenched legal ones.

Legal Background: Access to Evidence and Postconviction Relief

In order to understand the legal debate over postconviction DNA testing, a brief detour into criminal jurisprudence is necessary. The most important precedent was set in 1963 in *Brady v. Maryland*, when the Supreme Court ruled that a defendant in a criminal case is entitled to disclosure of any and all favorable and relevant evidence in the state's possession before trial. Before *Brady*, each side was free to withhold evidence from its opponent. *Brady* held that failure to disclose such evidence, irrespective of the motivations of the prosecutor, was a breach of due process.

Critical to the use of DNA testing in postconviction relief petitions is whether untested biological materials are subject to *Brady* guidelines. The current leading case on this issue is *Arizona v. Youngblood* (1988) in which the Supreme Court ruled that due process is violated only when the state fails to preserve, or destroys, evidence in "bad faith" (i.e., when the evidence could potentially exculpate a convict but is destroyed anyway). Thus, based on current constitutional doctrine, it is legal for prosecutors and law enforcement agents to destroy materials of no known exculpatory value, as long as that act does not violate any existing state or federal statute. Consequently, *Brady* established no clear right of postconviction access to biological material for DNA testing.

That said, lawyers have had some success advancing *Brady* arguments around the country, most notably in *Dabbs v. Vergari*.[3] In this 1990 New York case, an inmate sought DNA testing of physical evidence used to convict him of rape. The district attorney opposed this action, arguing that there was no statutory right to such a request. Finding in favor of the inmate, the court ruled that prosecutors should be held to the same standard to preserve and hand over exculpatory evidence to the defense both before and after trial. The court opined: "Due process is not a technical conception with a fixed content. . . . It is flexible and calls for such procedural protections as the particular situation demands. *Clearly, an advance in technology may constitute such a change in circumstance"* (*Dabbs* 1990, 768 [emphasis added]).

In a similar case in 2000, *Cherrix v. Braxton*, the court distinguished the inmate's claim from that in *Herrera* on the issue of evidence. Though the affidavits in *Herrera* "did not meet the standard of a truly persuasive showing of actual innocence," the court argued that "the circumstances in Cherrix's case are different. The evidence to be discovered in Cherrix's case constitutes DNA test results on seminal fluid seized from the body of the victim, which may be highly probative of the perpetrator's identity." The court then went on to state that the persuasiveness of DNA evidence on questions of guilt or innocence is "unquestionable" (*Cherrix* 2000, 767). Thus, at least a few state court judges have accepted the argument that DNA evidence is so powerful that it trumps not only the law's ordinary reliance on process as the guarantor of finality but also society's right to finality in criminal trials.

Postconviction Relief

Although a convicted prisoner can seek postconviction relief by several avenues, the most important is the writ of habeas corpus, which allows a prisoner to bring the authorities imprisoning him or her before a court of law to test the legality (constitutionality) of his conviction. In a series of cases over several decades, the U.S. Supreme Court established that the sole purpose of habeas corpus review is to test the constitutionality of a conviction, not to review its underlying factual basis. In other words, no matter how much a prisoner may wish to prove his or her innocence, the prisoner has no absolute right to do so after being convicted. Two recent developments in habeas law are especially relevant to the use of DNA evidence in postconviction relief: the Supreme Court's decisions in *Herrera* and *Schlup v. Delo* (1995); and the passage of the federal Antiterrorism and Effective Death Penalty Act of 1996 (AEDPA).[4]

Herrera, which involved the 1981 shooting deaths of two Texas highway patrolmen during a traffic stop, was decided just before postconviction DNA testing became an important part of the debate about the fairness and efficacy of the American criminal justice system. Leonel Herrera was arrested soon after the shootings based on a wide range of evidence, including eyewitness testimony, the fact that his girlfriend owned the car that had been stopped, serological data that matched blood on his pants to the one of the slain officers, as well as a handwritten note found in Herrera's pocket at the time of arrest strongly implying that he had committed the crime. In January 1982, Herrera was found guilty of murdering the second officer and was sentenced to death. Six months later, he pled guilty to the murder of the first officer, and unsuccessfully appealed the first conviction on the ground that some of the evidence was improperly admitted. He subsequently filed petitions for state and federal habeas corpus relief, both of which were denied.

More than eight years later, Herrera filed a second petition for state habeas corpus relief, and then for federal habeas relief, this time based on what he considered to be important new information not available at the first trial: two affidavits claiming that Herrera's now dead brother was the true perpetrator of the crimes. The District Court granted his request for a stay of execution so that this new evidence could be analyzed in court. On appeal, the U.S. Court of Appeals for the Fifth Circuit, vacated the stay, stating that the existence of newly discovered evidence relevant to the guilt of a state prisoner was not a ground for federal habeas corpus relief. Herrera appealed this judgment to the Supreme Court, which upheld the appellate decision.

In a 6–3 opinion, the Supreme Court held that a petitioner could launch a second petition for federal habeas corpus relief only if his constitutional claims were supplemented with a "colorable showing of factual innocence" (*Herrera* 1993, 400). No guidance was provided on exactly what such a showing might look like, but as a legal term "colorable" means plausible or believable. The Supreme Court also held that claims of actual innocence, in the absence of a constitutional claim, were not grounds for habeas corpus relief. Instead, they were merely "a gateway though which a habeas petitioner must pass to have his otherwise barred constitutional claim considered on the merits" (404). In a vigorous dissent, however, Justices Blackmun, Stevens, and Souter denounced this view, arguing that the execution of a person who has been validly convicted and sentenced, but who can prove his innocence with newly discovered evidence, was forbidden by the Eighth and Fourteenth Amendments (430–431).

At the heart of *Herrera* was a question of what to do with newly discovered evidence that could support or refute the validity of a guilty verdict. The *Herrera* majority held that for newly discovered evidence to lead to postconviction relief, it must reasonably have been unavailable at the initial trial, and it must also accompany a violation of constitutional rights. Thus, *Herrera* established that newly discovered evidence can matter only if it is linked to a constitutional violation. The main justifications for this conclusion were that "the passage of time only diminishes the reliability of criminal adjudications," and therefore that evidence based on affidavits alone evidence would not be powerful enough to guarantee a more exact finding of guilt or innocence if Herrera were to receive a new trial.[5] This view was codified by the passage of ADPEA in 1996, together with the requirement that habeas corpus relief must be applied for within one year after conviction in state court.

Schlup v. Delo

Two years after *Herrera*, the Supreme Court addressed the question of what should happen when a death row inmate who has exhausted all other avenues of postconviction relief claims actual innocence based on both new evidence and a constitutional violation at his original trial. In *Schlup*, the court held that when the two claims are made simultaneously by a death row inmate, the petitioner need show only that the constitutional error *probably* resulted in his wrongful conviction. In other words, he must convince the habeas court that "in light of the new evidence, it is more likely than not that no reasonable juror would have found him guilty beyond a reasonable doubt" (*Schlup* 1995, 200). Thus, the *Schlup* standard is slightly more lenient than that in *Herrera* (in that it is framed in the language of probabilities of innocence rather than certainty of innocence), but only because it seeks to prevent the most heinous miscarriage of justice—the execution of an innocent person.

Herrera and *Schlup* were both at play in *House v. Bell* (2006), in which the justices were asked to rule on what constitutes such a persuasive showing of actual innocence that it need not be accompanied by a constitutional claim to justify habeas relief. In this case, the Court directly addressed the issue of postconviction DNA testing for the first time. Tennessee death row inmate Paul House was seeking relief from his conviction and death sentence for the rape and murder of Carolyn Muncey, a woman who lived near him in a rural part of the state (Fisch 2006; Lane 2005).

During postconviction proceedings, House claimed that he had received ineffective counsel during his trial and presented three major pieces of new evidence that were not available when he was originally convicted. The first was the testimony of two women who claimed that Muncey's husband had confessed to killing his wife after they had been arguing. The second was evidence that the crime scene investigation was poorly handled and could have led to the spillage of Muncey's blood on House's clothing while the physical evidence was being transported from Tennessee to the FBI laboratory in Washington, D.C. Finally, new DNA tests showed that the semen on Muncey's clothing almost certainly belonged to her husband and not to House. If correct, this result meant that House most likely did not rape Muncey, taking away a crucial piece of evidence that linked him to the crime scene (Fisch 2006; Lane 2005).

After the Tennessee Supreme Court declined to grant relief, a federal District Court determined that of the three new pieces of evidence, only the DNA evidence was reliable. The court ultimately agreed with the state that the premeditated nature of House's crime (which was a centerpiece of the prosecution's case against him) was significantly more important to his conviction and death sentence than the suggestion that he was motivated by rape. Thus, although the DNA evidence certainly changed the case, it did not affect the guilty verdict. Neither the federal District Court nor any of the federal appellate courts that heard the case felt that this evidence was sufficient to establish his innocence, and none granted him a new trial.

House thereupon filed an appeal to the Supreme Court and the Court agreed to hear the case. During oral arguments on January 11, 2006, the justices focused heavily on how to weigh various forms of evidence. A key question was whether the new DNA evidence would have swayed the jury in the original trial either to declare House innocent or at least to not sentence him to death because of residual doubts about his guilt.[6] This issue was especially important for Breyer, who at one point put himself in the shoes of the jury and suggested that the confession evidence and the DNA evidence might have swayed him to a not guilty vote (oral argument in *House* 2006, 46–47). Scalia interjected that Breyer could not undertake such a thought experiment because he had no way of determining the credibility of the confession evidence, but Souter immediately argued that such an argument does not hold for DNA evidence. Any juror who heard the results of the DNA tests on the semen on Muncey's clothing, Souter observed, "would have to say that the only positive evidence that a rape was committed here would be evidence that pointed to the husband, not in fact to—to the defendant House" (49). Although there was some dispute among the lawyers and the justices about Souter's conclusion, several of the justices stated that the DNA evidence in the case at least called the stated motive for the crime (rape) into question (51–58). Thus, the debate among the justices was not about the absolute veracity of the DNA evidence, but only about its relevance to the total evidentiary picture of guilt and premeditation.

In a 5–3 ruling (Justice Alito was not yet a member when oral arguments were heard), the court determined that House's petition for postconviction relief was viable and granted him a new trial. Despite the heavy media focus on the DNA evidence in the months leading up to the oral arguments, the majority opinion, written by Justice Kennedy, did not single out DNA evidence as the determining factor in justifying relief. Instead, the majority argued that the three pieces of new evidence, taken in totality, suggested that "it is more likely than not that no reasonable juror viewing the record as a whole would lack reasonable doubt" about House's guilt (*House* 2006, 545). For the dissenters, this was the wrong standard to apply. In their view, new evidence must not merely cast doubt on House's conviction; it had to prove "that House was actually innocent, so that no reasonable juror would have convicted him in light of the new evidence" (548).

The Purification of DNA Profiling

The argument that a convicted felon has a constitutional right to DNA evidence even after he has exhausted all legal remedies rests squarely on the idea that DNA testing serves as a "truth machine" that can definitively determine guilt or innocence beyond doubt. Both Peter Neufeld, a noted liberal, and former U.S. Attorney General John Ashcroft, a noted conservative, have characterized the technique in this way (Neufeld 2003, 33; Ashcroft 2002). Since its first introduction as a forensic technique, DNA evidence has been endowed with almost mythic infallibility both by prosecutors using it to put defendants behind bars and by defense attorneys using it to free the wrongfully convicted from prison (Aronson 2007; Lynch et al. 2008). Perhaps the strongest claim made by the defense community in this regard can be found in *Actual Innocence*: "DNA testing is to justice what the telescope is for the stars: not a lesson in biochemistry, not a display of wonders of magnifying optical glass, but a way to see things as they really are. It is a revelation machine. And the evidence says that most likely, thousands of innocent people are in prison" (Scheck, Neufeld, and Dwyer 2000, xv).

The comparison of DNA testing to the telescope is revealing. As the historian of science Simon Schaffer has shown, when the telescope was introduced into astronomy, the visual data it produced were often highly ambiguous, leading to multiple interpretations among scientists. Further, many lay people simply did not trust an implement like the telescope to provide them with an accurate portrait of stars as they "really" were. Viewers had to be trained both to interpret the imperfect images created by the telescope and to believe that they actually represented reality (Schaffer 1983, 1989). In the same way, Scheck and Neufeld actively campaigned to convince judges, prosecutors, politicians, and the public that DNA was a revelation machine for exposing the faults of the criminal justice system that were not immediately obvious or apparent to most people (Scheck, Neufeld, and Dwyer 2000; Neufeld 2003).

Although Scheck and Neufeld's support was crucial in establishing the status of DNA profiling as the gold standard of forensic science, these two passionate advocates did not always have such a rosy view of the technique. Indeed, Scheck and Neufeld were responsible for generating significant controversy about the validity and reliability of DNA testing in the first few years after its introduction into the American legal system (Aronson 2007). Notable examples include *People v. Castro* (1989), in which they highlighted significant flaws in the laboratory procedure of one of the two private companies offering the technique; *United States v. Yee* (1991), in which they challenged the methods used by the FBI to calculate the probability of a false match between biological samples; and *People v. Orenthal James Simpson* (1994), in which they argued that although most of the technical problems associated with forensic DNA testing had been resolved, DNA evidence still could not automatically be trusted because of the fallibility and corruptibility of the human beings performing it (Scheck 2003; Thompson 1996). It should be noted that their strategy in the Simpson case was a marked departure from previous cases in which they sought to open up the "black box" of forensic DNA analysis to highlight its potential faults. By contrast, in the Simpson case, Scheck and Neufeld treated the laboratory technique almost as a black box and argued that the limiting factor was the skill, honesty, and integrity of the people responsible for managing the evidence. Indeed, summing up his attack on the evidence at the Simpson trial, Scheck declared, "garbage in, garbage out"—in other words, don't blame the technology when your inputs are fatally flawed (Lee and Tirnady 2003, 257–258). By switching the focus of attack to law enforcement officials—in this case, the discredited Los Angeles Police Department—Scheck offered in effect a preview of the strategic gear shift that led to the Innocence Project (Thompson 1996).

The Innocence Project

Following the Simpson trial, Scheck and Neufeld's mission became much bigger than protecting their legal clients from unreliable evidence. In 1992 they founded the Innocence Project, a nonprofit legal clinic at New York's Cardozo School of Law, where Scheck was a professor. The clinic was set up in order to free a few of what they believed were thousands of wrongfully convicted people languishing in American prisons. To succeed, however, they needed a form of proof that was so credible and convincing that prosecutors and law enforcement agents would be unable to disagree with them. They found this truth teller in DNA. Gone now were their one-time concerns about the integrity of forensic samples; gone (at least for a while) were their fears of lab error; and gone was their original skepticism toward scientific claims of infallibility. Scheck, Neufeld, and Dwyer's 2000 book, *Actual Innocence,* does not even mention their earlier experiences with DNA evidence. It is sanitized history, with DNA as the triumphant hero.

To be fair, Scheck and Neufeld had long argued that although DNA evidence was problematic when used for incrimination, it could be reliably used for exculpatory purposes, because no population genetics data were needed. A nonmatch requires no statistical interpretation. This view, however, ignored the ever present problems of contamination, degradation of forensic DNA samples, chain of custody issues, and lab misconduct. By carefully reviewing the circumstances of each case before accepting it, however, the Innocence Project rarely has to discuss these potential problems with DNA evidence publicly. The Innocence Project makes no secret of the fact that it accepts only those cases in which DNA evidence can yield conclusive proof of actual innocence.[7] In doing so, Innocence Project lawyers manage the image of DNA in the postconviction context so that there can be no question of its truth-telling power. The Innocence Project owes its success to this continuous purification of DNA profiling from its problematic social matrix.

A Fundamental Right to DNA Testing?

For many in the defense community, DNA profiling is so much stronger than other forms of evidence that it overrides traditional arguments about the sanctity of procedural finality in our legal system. If a DNA test can definitively adjudicate guilt or innocence, then it would be a constitutional violation, so the defense argument runs, to deny prisoners access to postconviction DNA testing. In a 2005 *Scientific American* article, Neufeld and Innocence Project policy analyst Sarah Tofte made exactly this case, arguing that "the dozens of DNA exonerations demonstrate that, a decade or more after conviction, DNA results are more reliable than eyewitnesses, confessions, and questionable forensic science introduced at the original trial . . . DNA, in limited situations, offers the criminal adjudicatory process a doctrine of certainty to replace the doctrine of finality" (Neufeld and Tofte 2005, 188–189).

However, because *Herrera* effectively blocked a prisoner's ability to obtain postconviction DNA testing to prove actual innocence, defense attorneys have had to pursue other legal avenues to gain access to biological materials for analysis (Neufeld and Tofte 2005, 189). In several cases,[8] the Innocence Project and affiliated attorneys made use of the 42 USC §1983 civil suit, which is a civil court action that allows a citizen to petition the federal government for relief or remedy when a state agent does not protect his or her constitutionally guaranteed rights (Vetter 2004). This legal mechanism was initially developed in 1871 in response to the failure of southern states to protect blacks from the Ku Klux Klan, but its use was expanded in the 1961 Supreme Court case *Monroe v. Pape* to provide federal remedies for state laws that were inadequate in theory or practice. In the criminal context, this may mean seeking compensation for unconstitutional treatment or demanding access to services or protections not provided by the state.

In *Harvey v. Horan* (2002a), James Harvey, a Virginia prisoner convicted of rape, sought a constitutional right of access to DNA evidence under §1983. This kind of challenge differs from a petition for habeas corpus in that a successful outcome neither secures the release nor proves the actual innocence of a convicted prisoner. At best, it can provide access to evidence that might establish actual innocence. Two central aspects of a §1983 suit are that the evidence is never automatically exculpatory, as test results could show that the DNA sample from the crime scene matches the plaintiff, and even if the evidence is exculpatory, the plaintiff must still file for habeas corpus or ask for a pardon in order to be released from prison. In other words, a §1983 suit must not seek to overturn a conviction, and it cannot be seen as bypassing state courts—it can only ask for evidence that the state is unwilling to hand over to the defendant for testing due to some legal or procedural defect.

Harvey originated in 1996, when the Innocence Project asked the Virginia Division of Forensic Science to hand over biological evidence for retesting. They asked again in 1998 and 1999, but their requests were denied. Harvey subsequently argued that the state's failure to test biological evidence using the latest Short Tandem Repeat (STR)-based DNA profiling technology violated his due process rights.[9] The federal district court hearing the case acknowledged such a right based on *Brady*, and also accepted that his claim was not for a writ of habeas corpus because he was not seeking immediate release from prison. Commonwealth Attorney Horan appealed this ruling.

The Court of Appeals for the Fourth Circuit reversed the district court decision, arguing in part that Harvey had not followed the proper procedure in making his claim for postconviction relief. The court stated that a prisoner could bring a §1983 claim only *after* the conviction or sentence is "reversed, expunged, invalidated, or impugned by the grant of a writ of habeas corpus" (Harvey 2002a, 374). As the Supreme Court had ruled in *Heck v. Humphrey* (1994), such civil suits could not be used to challenge a still valid criminal conviction (because there was no confirmation that any constitutional rights had been violated). The threat to finality loomed large in the Fourth Circuit court's thinking:

Harvey would have this court fashion a substantive right to postconviction DNA testing out of whole cloth or the vague contours of the Due Process Clause. We are asked to declare a general constitutional right for every inmate to continually challenge a valid conviction based on whatever technological advances may have occurred since his conviction became final. The Supreme Court has made clear that the finality of convictions cannot be brought into question by every change in the law. . . . Similarly, we believe that finality cannot be sacrificed to every change in technology. The possibility of postconviction developments, whether in law or science, is simply too great to justify judicially sanctioned constitutional attacks upon final criminal judgments. (*Harvey* 2002a, 375)

In other words, the court acknowledged that although finality is not a value that trumps all others, it can be overridden only in cases in which radically new evidence is discovered after trial.

In *Harvey*, the Fourth Circuit staked out a very conservative position with regard to the law's obligation to keep up with developing science and technology. As far as this court was concerned, the legal system has a valid, well-established mechanism for discovering the facts of a case that is not intrinsically inferior to scientific methods of truth making. In the interest of justice, already settled cases should not be reopened simply because some new scientific technique could potentially provide additional information not originally available at trial. In a society of seemingly continuous scientific change, doing so would mean that all judicial decisions would become provisional—never finished, always open to relitigation (*Harvey* 2002a, 375–376).

This argument implicitly denies the theory of the law lag—the idea that the legal system takes a long time to take notice of, understand, and come to grips with rapidly evolving science—*and* that it has an obligation to do better.[10] According to Fourth Circuit, the legal system has no duty to continually readjudicate old cases by the newest science; it must simply seek to ensure that the best available contemporary science is used at the time that the case is initially litigated. It is law, not science, that authorizes the final determination of guilt or innocence (*Harvey* 2002a, 376).

Obviously unhappy with the decision, the defense petitioned for rehearing and rehearing *en banc* (that is, with all members of the court sitting together) in March 2002. Both petitions were denied. Chief Judge Harvey Wilkinson filed an opinion supporting the denial, and Judge J. Michael Luttig filed an opinion against the decision (*Harvey*, denial of *en banc* motion). Luttig thereby became one of the few judges in the country to support a constitutional right to postconviction DNA testing—an unusual position for one of the most conservative jurists in the country. Until the appointment of Supreme Court Justices John Roberts and Samuel Alito, Luttig was on the Bush administration's short list of nominees, along with Wilkinson (Kirkpatrick 2005).

Wilkinson was clearly concerned that constitutionalizing a right to post-conviction access to DNA would foreclose broader democratic deliberation about the impact of science on society. In his view, it is not the court's prerogative to adapt established procedures to new scientific advances unless explicitly told to do so by Congress (*Harvey* 2002b, denial of *en banc* hearing, 301). Fairness and justice, he held, are guaranteed by already enshrined constitutional norms and the "orderly" processes set up to implement them. For Wilkinson, Harvey's §1983 suit was a blatant attempt to bypass Virginia's system of criminal justice and proceed directly into federal court. "Such disregard of process," he wrote, "is an anomaly in an area where criminal defendants, above all, rely on proper process to protect their rights. . . . Shorn of process, neither the innocent nor the public upon whom offenders prey will have any assurance of justice" (299).

Luttig found this reasoning faulty in light of the power of DNA-based technology to establish truth. He believed that the advances that led to DNA testing were "no ordinary developments, even for science." As a result, they could not be treated as "ordinary developments for law." Instead, they "must be recognized for the singularly significant developments that they are—in the class of cases for which they actually can prove factual innocence, the evidentiary equivalent of 'watershed' rules of constitutional law" (*Harvey* 2002b, 305–306).[11] After pointing out that the right to DNA testing must be tightly managed so as not to overwhelm the criminal justice system with spurious claims of innocence, Luttig stated that "it would be a high credit to our system of justice that it recognizes the need for, and imperative of, a safety valve in those rare instances where objective proof that the convicted actually did not commit the offense later becomes available through the progress of science" (306).[12] For Luttig, then, the law has an overriding duty to incorporate objective truth, and hence to defer to the exceptional truth-telling capability of DNA profiling.

These issues resurfaced in *Osborne v. District Attorney's Office* (2008), a §1983 case in the Court of Appeals for the Ninth Circuit.[13] William Osborne, a prisoner in Alaska, had been convicted of sexual assault and kidnapping, along with an accomplice. In 2002, he sought to compel the District Attorney's Office in Anchorage to allow him to test the biological evidence used to convict him in 1994 (a used condom and two hairs) with sophisticated DNA profiling techniques unavailable at the time of his original trial. Osborne argued, following the logic of *Dabbs* and subsequent cases, that the state's *Brady* obligations extend beyond well beyond the pretrial phase all the way through to the postconviction period, and further, that *Heck* is no barrier to a §1983 civil suit, because even if he gained access to evidence for further testing, it would not automatically invalidate his conviction. Following *Heck*, such determinations would have to be made in a separate criminal court proceeding (*Osborne* 2005a, 1056; *Osborne* 2008, 1122).

At the time, Alaska was one of three states that had no statute mandating postconviction access to biological evidence, so the state Court of Appeals heard such cases (a postconviction DNA testing statute was ultimately passed in May 2010). In Osborne's case, the court was "reluctant to hold that Alaska law offers no remedy to defendants who could provide their factual innocence," and therefore devised a three-part test for access to biological evidence based on policies deemed to exist in other states (*Osborne* 2005b, 995). In order to gain access to biological evidence, Osborne's request had to satisfy the following criteria: (1) the original conviction had to rest primarily on eyewitness identification, (2) there had to be doubt in the identification of Osborne by the witness, and (3) any evidence produced had to be conclusively exculpatory. Osborne did not pass this test because numerous other forms of evidence were presented at trial, including a gun found in Osborne's car that matched shell casings recovered by police at the crime scene. He was therefore denied relief.

After several rounds of litigation, Osborne's legal team persuaded the Anchorage Police Department to hand over evidence for testing. Although the district attorney opposed this action, the district court in Anchorage and the Ninth Circuit Court of Appeals decided in favor of Osborne, finding (on the basis of Luttig's reasoning) that he had a "very limited constitutional right to the testing sought" based on the novelty and revelatory potential of the evidence and the fact that it would not be directly used to invalidate his conviction (*Osborne* 2006, 1080–1081). Unsatisfied with this result, the district attorney asked the U.S. Supreme Court to review the decision of the Ninth Circuit, setting up a definitive test of the existence of a constitutional right to postconviction DNA testing.

The Supreme Court reversed the Ninth Circuit, denying that any due process violation occurred. Alaska, the court noted, had a legitimate framework for postconviction relief that, however imperfect, did not "offend" fundamental principles of justice or fairness (*District Attorney's Office v. Osborne* 2009, 16). This, in the majority's view, would be the only basis for upsetting a state's postconviction relief procedures. More to the point, the Supreme Court also rejected Osborne's plea to recognize a freestanding right to DNA testing in the absence of some constitutional error at trial, because they felt this was a matter for legislatures to deal with (1–2). In the majority's view, articulated by Chief Justice Roberts, "there is no reason to constitutionalize the issue in this way" (2). Relying on Wilkinson's reasoning in *Harvey*, Roberts argued that "the availability of technologies not available at trial cannot mean that every criminal conviction, or even every criminal conviction involving biological evidence, is suddenly in doubt. The dilemma is how to harness DNA's power to prove innocence without unnecessarily overthrowing the established system of criminal justice" (8). Although the four dissenting justices argued that the benefits of absolute certainty provided by this revolutionary new technique outweighed the risks associated with violating the established legal order, the majority countered that "there is no long history of such a right and the mere novelty of such a claim is reason enough to doubt that substantive due process sustains it" (19).

The majority also took aim at the very notion that served as the foundation for calls to constitutionalize a right to postconviction DNA testing—that modern STR analysis can provide conclusive proof of guilt or innocence in an efficient, low-cost manner. Taking a page from Scheck and Neufeld's old playbook, the majority adopted a skeptical view of the certainty accorded to DNA profiling. Contesting the dissenting judges' claim that "the DNA test Osborne seeks is a simple one, its costs modest, and its results uniquely precise" (*District Attorney's Office v. Osborne* [2009],

Stevens, J. dissent, 1), the majority quoted at length from law professor Erin Murphy's *Emory Law Journal* article on "the subjectivity inherent in forensic DNA typing" (Murphy 2008). There, Murphy reopened the black box that Scheck and Neufeld had fought to close, highlighting such issues as contamination, degradation of DNA, and the ambiguity of forensic samples containing biological material from multiple people. Persuaded by her deconstruction, the court also went along with her caution against overconfidence in the results of DNA testing.

Conclusion

The Supreme Court in *District Attorney's Office v. Osborne* (2009) seemed to recognize, and capitalize upon, the central point of the coproduction framework: that the construction of a constitutional right to DNA testing is intimately linked to the construction of DNA testing itself as a foolproof and fail-safe technology. The technique's status was achieved as much through social action as through scientific advance, and the legal system ought not to treat it as an intrinsically infallible "revelation machine" (Aronson 2007; Jasanoff 2006; Lynch et al. 2008). Further, crafting a constitutional right to access to DNA evidence would not have greatly advanced the cause of justice in the long term. With each passing year, the subset of cases in which previously untested genetic evidence exists grows smaller and smaller. In fact, within a relatively short time, the era of DNA-based exonerations may well be over. Even during this transitional era, in the vast majority of postconviction cases, there is simply no biological evidence available for testing when a prisoner claims innocence.

Yet the failure to construct a constitutional right to postconviction DNA testing does not mean that Luttig's call for a "safety valve" for the wrongfully convicted was unfounded. Indeed, the justification for postconviction access to strong evidence of innocence is so abundantly clear (thanks, in large part, to the excellent but not infallible technology of DNA profiling) that it need not depend on any exaggerated belief in the infallibility of science. The way forward is to argue for a constitutional right to any evidence that meets the *Schlup* standard, regardless of whether the case involves a capital crime. Thus, any new evidence making it "more likely than not" that a reasonable juror, presented with that evidence, could not have convicted the defendant would potentially trigger a reexamination of the conviction. Under such a rule, convicted felons could gain access not only to DNA evidence in cases where it might have a major impact on the outcome, but also to any other form of evidence that might profoundly affect a jury's decision. It would, of course, be left to the courts to determine, state by state, what kinds of evidence would pass this test in individual cases, but state courts already play this gate-keeping role in criminal trials.

We may give Luttig the last word with respect to the ideal relationship between legal procedure and scientific techniques—or between social finality and epistemic certainty—in meting out justice. In his criticism of the decision not to rehear Harvey's postconviction case, Luttig wrote that if "it is agreed that, in a given class of cases, it would be possible to establish to a certainty through such further analysis that one did not in fact commit the crime for which he was convicted and sentenced, then grave harm would come to the Constitution were it to be dismissively interpreted as foreclosing access to such evidence under any and all circumstances and for any and all purposes (judicial or even executive). The Constitution is not so static" (*Harvey* 2002b, 306).

In the end, we can conclude that both finality (just process) and certainty (DNA typing) are important social achievements. In addition to safeguarding the rights of the defendants and preserving social order, the legal must system should endeavor to ensure that neither value gets elevated to the status of a false god.

Notes

1. See the Innocence Project's website for the latest total: <http://www.innocence project.org>.

2. "Purification" here refers to the process by which the social dimensions of science are hidden in order to ensure that results are seen by outsiders to be unmediated representations of nature as it really is. Purification thus renders contingency, human intervention, and other sources of error invisible to the untrained eye (Latour 1993).

3. See also *State v. Thomas* (1990) and *Sewell v. State* (1992).

4. Habeas corpus was first codified in England in a 1641 act that specifically allowed the courts of the King's Bench or common pleas to examine the legality of a detention (see Capra and Saltzburg 2007).

5. See also: *Ford v. Wainright* (1986) and *Johnson v. Mississippi* (1988).

6. This action is consistent with the empiricist behavior of judges described by Sheila Jasanoff (Jasanoff 1998, 2002).

7. See Innocence Project, "FAQ: How do you choose your cases? How many letters do you receive?" (July 6, 2010) at <http://www.innocenceproject.org/Content /103.php>.

8. *Harvey v. Horan* (2002a); *Godschalk v. Montgomery County District Attorney's Office and Bruce Castor* (2001); *Bradley v. Pryor* (2002); *McKithen v. Brown*

(2007); *Breest v. N.H. Attorney General* (2008); and *Osborne v. District Attorney's Office* (2006)

9. In oral arguments, the defense conceded that Harvey received due process under both law and science when he was convicted in 1990.

10. See Jasanoff, chapter 1, this volume.

11. See *Teague v. Lane* (1989).

12. Based on Luttig's logic, several courts have granted access to postconviction DNA evidence, but none of these cases led to a substantial shift in the legal landscape. See *Bradley v. Pryor* (2002), *Breest v. N.H. Attorney General* (2008), and *McKithen v. Brown* (2007).

13. There are several related cases in this matter with the same or similar names. See the Cases Cited section for details.

References

Aronson, Jay D. 2007. *Genetic Witness: Science, Law, and Controversy in the Making of DNA Profiling*. New Brunswick, N.J.: Rutgers University Press.

Aronson, Jay D. and Simon A. Cole. 2009. Science and the Death Penalty: DNA, Innocence, and the Debate over Capital Punishment in the United States. *Law and Social Inquiry* 34 (3): 603–633.

Ashcroft, John. 2002. *Attorney General Transcript: News Conference—DNA Initiative*, March 4. Retrieved from <http://www.usdoj.gov/archive/ag/speeches/2002/030402newsconferncednainitiative.htm> (accessed July 12, 2010).

Bedau, Hugo Adam, and Michael L. Radelet. 1987. Miscarriages of Justice in Potentially Capital Cases. *Stanford Law Review* 40:21–179.

Berger, Margaret A. 2004. Lessons from DNA: Restriking the Balance between Finality and Justice. In *DNA and the Criminal Justice System: The Technology of Justice*, ed. D. Lazer, 109–131. Cambridge, Mass.: MIT Press.

Blackstone, William. 1765–1769. *Commentaries on the Laws of England. Book Four: Of Public Wrongs*. Oxford: Clarendon Press.

Borchard, Edwin. 1932. *Convicting the Innocent: Sixty-five Actual Errors of Criminal Justice*. Garden City, N.Y.: Garden City Publishing Co.

Capra, Daniel J., and Stephen A. Saltzburg. 2007. *American Criminal Procedure, Cases and Commentary*. 8th ed. Eagan, Minn.: West.

Fisch, Daniel. 2006. *liibulletin: Supreme Court Oral Argument Previews: House v. Bell (04–8990)*. Cornell Law School Legal Information Institute 2006 [cited April 3].

Gross, Samuel R., Kristen Jacoby, Daniel J. Matheson, Nicholas Montgomery, and Sujata Patil. 2005. Exonerations in the United States 1989 through 2003. *Journal of Criminal Law & Criminology* 95:523–560.

Innocence Project. 2010. Access to Post-Conviction DNA Testing, July 12. Retrieved from <http://www.innocenceproject.org/Content/304.php> (accessed July 12, 2010).

Jasanoff, Sheila. 1998. The Eye of Everyman: Witnessing DNA in the Simpson Trial. *Social Studies of Science* 28 (5–6): 713–740.

Jasanoff, Sheila. 2002. Science and the Statistical Victim: Modernizing Knowledge in the Breast Implant Litigation. *Social Studies of Science* 32 (1): 37–69.

Jasanoff, Sheila. 2006. Just Evidence: The Limits of Science in the Legal Process. *Journal of Law, Medicine & Ethics* 34 (2): 328–341.

Kirkpatrick, David. 2005. Senate Democrats Are Shifting Focus from Roberts to Other Seat. *New York Times*, September 9, A1.

Lane, Charles. 2005. Court May Revise Rule on Death Row Appeals. *Washington Post*, June 29, A3.

Latour, Bruno. 1993. *We Have Never Been Modern*. Cambridge, Mass.: Harvard University Press.

Leahy, Patrick. 2001. Statement from Hearing on Post-Conviction DNA Testing: When Is Justice Served? United States Senate Committee on the Judiciary, June 13, 2000. Retrieved from <http://ftp.resource.org/gpo.gov/hearings/106s/74753.pdf>.

Lee, Henry C., and Frank Tirnady. 2003. *Blood Evidence: How DNA Is Revolutionizing the Way We Solve Crimes*. New York: Basic Books.

Lynch, Michael, Simon Cole, Ruth McNally, and Kathleen Jordan. 2008. *Truth Machine: The Contentious History of DNA Fingerprinting*. Chicago: University of Chicago Press.

Murphy, Erin Elizabeth. 2008. The Art in the Science of DNA: A Layperson's Guide to the Subjectivity Inherent in Forensic DNA Typing. *Emory Law Journal* 58:489–512.

Neufeld, Peter J. 2003. Statement from Advancing Justice Through DNA Technology: Hearing before Subcommittee on Crime, Terrorism and Homeland Security, House Judiciary Committee, July 17, 2003. Retrieved from <http://judiciary.house.gov/legacy/88394.pdf>.

Neufeld, Peter J., and Sarah Tofte. 2005. A Fundamental Right to Post-conviction DNA Testing. In *Rights and Liberties in the Biotech Age*, ed. S. Krimsky and P. Shorett, 185–193. Lanham, Md.: Rowman and Littlefield.

Oral Argument in Paul Gregory House v. Ricky Bell, Warden (U.S. Supreme Court, No. 04–8990), January 11, 2006.

Radelet, Michael L., Hugo Adam Bedau, and Constance L. Putnam. 1992. *In Spite of Innocence: Erroneous Convictions in Capital Cases*. Boston: Northeastern University Press.

Radin, Edward. 1964. *The Innocents*. New York: William Morrow.

Schaffer, Simon. 1983. Natural Philosophy and Public Spectacle in the Eighteenth Century. *History of Science* 21:1–43.

Schaffer, Simon. 1989. The Nebular Hypothesis and the Science of Progress. In *History, Humanity, and Evolution*, ed. J. R. Moore, 131–164. Cambridge: Cambridge University Press.

Scheck, Barry. 2003. Interview with Harry Kriesler as part of "Conversations with History Series at UC-Berkeley," July 25. Retrieved from <http://globetrotter

.berkeley.edu/people3/Scheck/scheck-con0.html>, 3 (accessed November 29, 2010).

Scheck, Barry, Peter Neufeld, and Jim Dwyer. 2000. *Actual Innocence: Five Days to Execution, and Other Dispatches from the Wrongly Convicted.* New York: Doubleday.

Thompson, William C. 1996. DNA Evidence in the OJ Simpson Trial. *University of Colorado Law Review. University of Colorado (Boulder Campus). School of Law* 67 (4): 827–857.

Vetter, Benjamin. 2004. Habeas, Section 1983, and Post-Conviction Access to DNA Evidence. *University of Chicago Law Review* 71:587–615.

Cases Cited

Arizona v. Youngblood, 488 U.S. 51 (U.S. Supreme Court, 1988).

Bradley v. Pryor, 305 F.3d 1287 (11th Circuit 2002).

Brady v. Maryland, 373 U.S. 83 (U.S. Supreme Court, 1963).

Breest v. N.H. Attorney General, CV-06–361-SM (U.S. Dist. Ct NH, January 18, 2008).

Cherrix v. Braxton, 131 F.Supp.2d 756 (U.S. District Court, Eastern District Virginia, 2000; referred to as *Cherrix* 2000).

Dabbs v. Vergari, 570 N.Y.S.2d 765 (Superior Court of New York, 1990; referred to as *Dabbs* 1990).

District Attorney's Office for the Third Judicial District, et al. v. Osborne, 517 U.S. ___ (U.S. Supreme Court, 2009; referred to as *District Attorney's Office v. Osborne* 2009) (note that all page numbers in this article refer to the unpublished [slip] opinion).

Ford v. Wainright, 477 U.S. 399 (U.S. Supreme Court, 1986).

Godschalk v. Montgomery County District Attorney's Office and Bruce Castor, Civil Action 00–9535, 2001 (2001 WL 1159857).

Harvey v. Horan, 278 F.3d 370 (Fourth Circuit Court of Appeals, 2002; referred to as *Harvey* 2002a).

Harvey v. Horan, denial of *en banc* hearing, 285 F.3d 298 (Fourth Circuit Court of Appeals, 2002; referred to as *Harvey* 2002b).

Heck v. Humphrey, 512 U.S. 477 (U.S. Supreme Court, 1994).

Herrera v. Collins, 506 U.S. 390 (U.S. Supreme Court, 1993; referred to as *Herrera* 1993).

House v. Bell, 547 US 518 (U.S. Supreme Court, 2006; referred to as *House* 2006).

Johnson v. Mississippi, 486 U.S. 578 (U.S. Supreme Court, 1988).

Mackey v. United States, 401 U.S. 667, 691 (U.S. Supreme Court, 1971).

McKithen v. Brown, 481 F.3d 89 (2d Cir. 2007).

Monroe v. Pape, 365 U.S. 167 (U.S. Supreme Court, 1961).

Osborne v. District Attorney's Office, 423 F.3d 1050 (Ninth Circuit Court of Appeals, 2005; referred to as *Osborne* 2005a).

Osborne v. State, 110 P.3d 986 (Alaska Court of Appeals, 2005; referred to as *Osborne* 2005b).

Osborne v. District Attorney's Office, 445 F. Supp. 2d 1079 (Anchorage District Court 2006; referred to as *Osborne* 2006).

Osborne v. District Attorney's Office, 521 F.3d 118 (Ninth Circuit Court of Appeals, 2008; referred to as *Osborne* 2008).

People v. Castro, 545 N.Y.S.2d 985 (NY Superior Court, 1989).

People v. Orenthal James Simpson, Case No. BA097211 (LA County Superior Court, October 4, 1994).

Sewell v. State, 592 N.E.2d 705 (Indiana Court of Appeals, 1992).

Schlup v. Delo, 513 U.S. 298 (U.S. Supreme Court, 1995; referred to as *Schlup* 1995).

State v. Thomas, 586 A.2d 250 (Sup. Ct. NJ, 1990).

Teague v. Lane, 489 U.S. 288 (U.S. Supreme Court, 1989).

United States v. Yee, 134 F.R.D 161 (ND Ohio 1991).

7

Judicial Imaginaries of Technology: Constitutional Law and the Forensic DNA Databases

David E. Winickoff

The question we confront today is what limits there are upon this power of technology to shrink the realm of guaranteed privacy.
—Justice Antonin Scalia, *Kyllo v. United States* (2001)

As two powerful epistemic institutions, law and technoscience work together in sustaining models of the individual and society and in producing ruling classifications and categories (Jasanoff 2008). An important case in point concerns new technologies of surveillance and their encounters with the Fourth Amendment of the U.S. Constitution. This "right of the people to be secure in their persons, houses, papers, and effects, against unreasonable searches and seizures" has steadily evolved through confrontations between civil liberties claims and advances in wiretapping, aerial photography, and increasingly sensitive microphones (Power 1989). Although STS scholars have closely tracked the bumpy process of bringing forensic DNA evidence into the courtroom (Aronson 2006; Lynch et al. 2008), they have not yet addressed the constitutional review of DNA collection for use in large-scale searchable databases. Merging the biological and informational, forensic DNA databases are reshaping legal understandings of security, freedom, and identity. These evolving tools are also restructuring the relations of criminal bodies and bodies politic. In short, they are deeply implicated in what this volume identifies as the evolving bioconstitutional order.

Forensic DNA databases have expanded considerably in size and scope since their inception in the early 1990s. Individual U.S. states began enacting forensic DNA data banking statutes before football star O. J. Simpson and DNA analyses of his blood infamously went to trial in 1994, and before admissibility standards for DNA evidence became stabilized. By 1999, all fifty states had enacted statutes providing for the mandatory DNA banking of blood or saliva samples from those convicted of certain felonies (Bieber 2004). Though most of these databases were originally slated to include only persons convicted of serious sexual offenses and violent crimes, many have expanded over the years to include all convicted felons and even all arrestees: as of September 2009, fourteen states were collecting samples for their databases from many people who are merely arrested (Biancamano 2009). The U.S. government has begun collecting DNA samples from all citizens arrested in connection with any federal crime and from many immigrants held by federal authorities, which is likely to add more than one million individuals per year to the federal database (Nakashima and Hsu 2008). These state and federal databases are connected through a digital network coordinated by the FBI. This network, called CODIS (Combined DNA Index System), enables federal, state, and local crime labs to exchange and compare DNA profiles electronically, thereby linking samples found at crime scenes to other samples and, by extension, to the individuals whose samples are in the database.

Although many new forms of technological surveillance and restraint have tended to evade significant judicial review (Murphy 2008), this is not the case for large-scale forensic DNA databases. Since the 1992 case of *Jones v. Murray*, judges in the federal circuit courts of appeal have confronted Fourth Amendment challenges to the system of searches and seizures authorized under DNA database legislation and the CODIS system. These encounters have not only unsettled, but actively reconfigured, the scope of civil liberty and the doctrinal architecture of the Fourth Amendment. None of these Fourth Amendment challenges at the circuit level has succeeded, but the doctrinal dispensation of the cases has varied widely. This variability presents a good opportunity for comparison. Previous work has shown how theories and narratives of technology shape public decision making and cultural history (Wynne 1988; Callon 1987; Hughes 2004). But how do new biotechnological imaginations work their way into the legal and social order? A detailed focus on case law will be necessary to answer this question, precisely because a bioconstitutionalist research design rejects hasty generalizations and asks for attention to details, translations, and transformations.

The cases discussed here illustrate how judges determine what constitutes due process against tacit and explicit background understandings of technological risk. The decisions deploy different Fourth Amendment doctrines in effect as a form of risk management in order to control the imagined hazards of these forensic searches in silico. Constitutional legitimization of the new DNA databases thus has depended upon what I call "judicial imaginaries of technology"—including tacit analogies, framing, models of social adoption of technology, and risk assessment. Technological imaginaries are at once ontologies, theories, sociologies, and narratives of technoscience that enable and construct social order (Jasanoff and Kim 2009). In the Fourth Amendment cases, the technological imaginaries of judges condition their doctrinal choices and the terms of technological adoption even as they help produce new categories of criminal subjects.

The Fourth Amendment

Infrared goggles, wiretaps, electronic bugs: these are the mainstays of Fourth Amendment doctrine as found in first-year law textbooks. It is here that judges consider whether new surveillance technologies violate the traditional Constitutional rubric: "the right of the people to be secure in their persons, houses, papers, and effects, against unreasonable searches and seizures, shall not be violated, and no Warrants shall issue, but upon probable cause, supported by Oath or affirmation, and particularly describing the place to be searched, and the persons or things seized."

Throughout the recent history of the Fourth Amendment, judges have struggled to define the boundaries of privacy in the face of changing technological and social norms. In *Katz v. United States*, the seminal case in modern search and seizure law, the court held that a police activity requires a warrant, or a special exception to a warrant, if it occurs in a place where the person had a "reasonable expectation of privacy."[1] Jurists have pointed out that privacy may tend to erode if reasonable expectations change, as indeed they have eroded in the era of information explosion.

In particular, since 1992, courts have refereed the confrontation of CODIS and the Fourth Amendment. As a result, through inevitable processes of normative and epistemic coproduction, novel conceptions of privacy and new criminal kinds, or definitions of what constitutes a criminal, have emerged. By allowing law enforcement officers to search existing DNA samples left at crime scenes against a large database of profiles, CODIS promises an efficient way of connecting individuals to crimes without a physical roundup or questioning. In the eyes of police and prosecutors, it amounts to a powerful law enforcement tool, a huge roundup at a click whose intrusiveness is minimal, electronic, and virtual. The legal claim made by convicts, parolees, and others included or slated for inclusion in a forensic database is that the new forensic DNA statutes mandate activities that constitute an unreasonable search or seizure under the Fourth Amendment. Such activity includes blood draws or other sampling without consent, banking of samples, creation of a genetic profile based on the sample, and inclusion of profiles within a network subject to repeated electronic scans.

None of the activities cited in the Fourth Amendment cases were being pursued with search warrants. Assuming these activities would be deemed "searches" under the amendment, would they then qualify for one of the traditional exceptions to the rule? This question has not been consistently answered by the courts, not just because of different conceptions of the technology but also because of disagreements about the scope of privacy in the face of police power.

The first clause of the Fourth Amendment is referred to as the "reasonableness" clause. Its exact relationship to the "warrant clause" is not made clear in the amendment itself, which has raised a disagreement in interpretation. Painting with broad strokes, there are two competing theories about how the two clauses relate (Bradley 1993). One theory holds that the warrant clause defines and gives meaning to the reasonableness clause (Maclin 1994, 33). According to this view, a warrant is required for every search and seizure, as long as it is "practicable" to obtain one. This view depends on what the Supreme Court has described as the "cardinal principle"[2] of the Fourth Amendment, namely that "searches conducted outside the judicial process, without prior approval by judge or magistrate, are per se unreasonable under the Fourth Amendment—subject only to a few specifically established and well-delineated exceptions."[3] The second and opposing viewpoint emphasizes that the Constitution's preference for warrants is a judicial construct rather than a textual requirement.[4] Advocates of this model contend that the warrant clause does not and should not inform the reasonableness clause. In this view, the reasonableness clause encompasses all searches and seizures, and requires only that such searches and seizures be "reasonable," taking into consideration all the relevant factors.

One issue not in dispute is that the doctrine of "probable cause" serves to mediate between these two positions. The existence of a "cause" that justifies the issue of a warrant disciplines both the issuing judge and the searching authority. And regardless of whether a warrant has been issued for a search of a "protected zone" of privacy, searches are almost always deemed unreasonable unless "the facts and circumstances" known by the searcher "are sufficient in themselves to warrant a person of reasonable caution in the belief" that the search will lead to evidence of a crime.

The paradigmatic rule *seems* to be one of requiring a warrant, and that a search without a warrant is deemed intrinsically unreasonable and unconstitutional unless one of the exceptions to the warrant requirement is demonstrated. Another way of stating this is that though there is no warrant *requirement*, there is a constitutional *preference* for warrants. Exceptions exist, but a showing of "probable cause" or "reasonable suspicion" is needed to sustain the constitutionality of searches that seek to invoke one of the exceptions. Only one warrant exception does not require a showing of some sort of individualized suspicion, and that is the "special needs" exception that is invoked by some of the DNA database cases detailed shortly.

Indeterminacy in the Courts

Judges have disagreed about how forensic DNA databases engage with the doctrinal framework of the Fourth Amendment. Although courts have faced the same constitutional questions—similar statutes and nearly identical legal precedents—they have proposed three different doctrinal classifications in deciding these cases: *reasonableness, special needs*, and *individualized suspicion*. In short, there has been strong judicial indeterminacy, pointing toward divergent judicial imaginaries. The three positions are illustrated by leading cases from three federal circuits: *Jones v. Murray* (4th Cir. 1992), *Roe v. Marcotte* (2nd Cir. 1999), and *U.S. v. Kincade* (9th Cir. 2003), reheard *en banc*[5] in 2004 (referred to as *Kincade II* in what follows).

"Totality of the Circumstances"

The earliest federal appellate case to consider the Fourth Amendment constitutionality of a forensic DNA database statute is *Jones v. Murray*[6] in 1992, and this case became the leading exemplar of the so-called reasonableness or totality of the circumstances approach. Most other courts have taken this approach and found that forensic DNA statutes survive constitutional scrutiny. In *Jones*, six inmates from the Tazewell Correctional Unit Number 31 challenged the Virginia DNA database legislation requiring convicted felons to submit blood samples for DNA analysis. The inmates argued that coerced extraction of blood violated the rule against unreasonable searches and seizures. The court upheld the district court's view that the DNA statute did not violate the Fourth Amendment rights of the inmates.

True to the traditional doctrinal scheme of such inquiries, the court first considered whether this governmental activity intruded upon a traditionally protected zone of privacy, a question that it answered affirmatively. Writing for the majority, Judge Niemeyer stated that "the bodily intrusion resulting from taking a blood sample constitutes a search within the scope of the Fourth Amendment" (*Jones*, 306). The inmates argued that *all* governmental searches conducted in the context of criminal law enforcement require a warrant, or at least some sort of "individualized suspicion," but the court rejected the application of such a rule to the case at hand. They had "not been made aware of any case establishing a *per se*" rule requiring probable cause for a "limited search" conducted "for the purpose of ascertaining and recording the identity of a person who is lawfully confined to a prison" (306).

The court reasoned that "probable cause" supplied the basis for bringing the person within the criminal justice system, and "with the person's loss of liberty upon arrest comes the loss of at least some, if not all, rights to personal privacy otherwise protected by the Fourth Amendment." The court next stated another general rule that "when a suspect is arrested upon probable cause, his identification becomes a matter of legitimate state interest and he can hardly claim privacy in it" (306). Citing the example of fingerprint and other "booking" procedures for every suspect arrested for a felony, the court wrote that "the identification of suspects is relevant not only to solving the crime for which the suspect is arrested, but also for maintaining a permanent record to solve other past and future crimes" (306).

The court next went directly to the "balancing test" required of many reasonableness judgments in the law, weighing the significance of the intrusion to privacy against the government's interest "in preserving a permanent identification record of convicted felons for resolving past and future crimes" (*Jones*, 307). In order to do this, however, the court had to justify its departure from the typical Fourth Amendment scheme of a per se requirement of at least "probable cause" for a particular crime. Accordingly, in footnote 2, the court explained: "Because we consider the cases which involve the Fourth Amendment rights of prison inmates to comprise a separate category of cases to which the usual per se requirement of probable cause does not apply, there is no cause to address whether the so-called "special needs" exception, relied on by the district court applies in this case" (307n2). In other words, the doctrinal choice to bypass the warrant requirement turns in part on the argument that prison inmates as a class have diminished rights in comparison to those of free and innocent citizens.

The court seems to suggest that arrestees held with "probable cause" forfeit any privacy interest in their "identification," which encompasses the inclusion of DNA samples in databases. Furthermore, prison inmates are considered a class of persons for whom finding individual probable cause of new crimes is not even required: their Fourth Amendment privacy has been diminished upon conviction and confinement. Here, the implication is that the government interest in maintaining the database outweighs the privacy interests of convicted criminals. Accordingly, the intrusion to privacy and the government's interests can be balanced and weighed without imposing the mechanisms of the probable cause or warrant requirement. In *Jones*, the court concluded that the level of intrusion was minimal, whereas the state's interests were significant. Many courts since *Jones* have used this reasoning, and the balancing has always come out the same way.[7] In these cases, prisoners constitute a de facto permanent class of "usual suspects."

Jones v. Murray is of interest not only for its majority opinion, but also for a strongly worded dissent from Judge Murnaghan. Because this dissent shaped how subsequent circuit courts have addressed these cases, it is important to review his position. Judge Murnaghan concurred with the *Jones* majority to the extent that it upheld the Virginia statute "as applied to violent felons," but he dissented from "the majority's determination of the constitutionality of the statute as applied to prisoners convicted of non-violent crimes" (*Jones*, 311). Judge Murnaghan argued that although precedent established that prisoners "give up specific aspects of their reasonable expectation of privacy," these forfeitures have been necessitated by "practical concerns relating to living conditions" and "ensuring prison security" (312). These are limited exceptions to a privacy rule that protects prisoners as much as free citizens against unjustified intrusions: "Prisoners do not lose an expectation of privacy with regard to blood testing, and the Commonwealth's articulated interest in the testing of non-violent felons does not counter-balance the privacy involved in the procedure" (311).

Harkening back to concerns about generalized and overbroad search warrants, he states that "there exists no blanket authorization of searches involving intrusions under the skin, for which no individual, whether in prison or out, loses a reasonable expectation" (*Jones*, 311). Murnaghan recognizes that the majority's opinion ratifies systematic deprivation of a Fourth Amendment expectation of privacy of those in prison.

In making a substantive distinction between the constitutionality of searches with respect to violent versus nonviolent offenders, Judge Murnaghan relied heavily on statistical studies of criminal recidivism. Murnaghan cited a Virginia governmental report that came into the record, the Report of the Joint Subcommittee Studying Creation of a DNA Test Bank. He noted that the Report concluded only that "the recidivism data supported the inclusion of plaintiffs convicted for felony sex offenses, assault, capital murder, first and second degree murder, voluntary manslaughter, larceny and burglary" (314).

Nonetheless, the report recommended the testing of *all* remaining felons, not because such testing would be likely to help solve crimes, but only because their inclusion would make the data bank "more efficient and cost effective" (Virginia Joint Subcommittee 1990). Judge Murnaghan pointed out that a similar rationale could be used to justify the inclusion of any citizen if it lessened the state's workload.

"Special Needs"

When the Second Circuit Court encountered the virtual roundup in 1999, Judge Murnaghan's dissent in *Jones* helped convince the court to choose a different theory of constitutionality for the DNA databases. In the case of *Roe v. Marcotte*, a group of convicted sexual offenders challenged the constitutionality of a Connecticut statute that, among other things, required all convicted sexual offenders to submit a blood sample for analysis and inclusion in the state DNA databank. Relying on arguments set out in Murnaghan's dissent, the court came to the same result as *Jones* but used a different legal doctrine.

Writing for the unanimous three-member panel in *Roe v. Marcotte*, Judge Poole declared that Judge Murnaghan's analysis "provides a more compelling rationale for upholding the DNA statute's constitutionality than does the *Jones* majority opinion."[8] Indeed, Judge Poole repeated Murnaghan's acerbic critique of the *Jones* majority's "strikingly truncated view of the Fourth Amendment protections afforded to a convicted felon" (*Roe*, 81).

The Second Circuit shared Murnaghan's concern at least enough to require a finding that the case at hand fall within an established exception to a per se rule of probable cause. The court recites the rule that "in general, searches performed in the absence of a warrant and pursuant to an exception must nevertheless be predicated upon 'probable cause to believe that the person to be searched has violated the law,' or at the very least, 'some quantum of individualized suspicion.'"[9] Rather than moving directly ahead to a balancing test, as the *Jones* court did on the idea that all prisoners lose their "expectation of privacy," the *Roe* court attempted to preserve the presumption against warrantless and suspicionless searches, even for incarcerated prisoners.

The *Roe* court explained that an exception to this "individualized suspicion" rule applicable in the case at hand was the so-called special needs doctrine, which states: "In certain circumstances, generally outside the traditional law enforcement setting, a search may be reasonable even when predicated upon less than probable cause or individualized suspicion where 'special needs, beyond the normal need for law enforcement render those requirements impracticable'" (*Roe*, 77). If courts determine that the searches in question qualify under this exception, then and only then may they proceed to a balancing analysis for "reasonableness."

But this doctrine is reserved for those "special needs, beyond the normal need for law enforcement." In order to make the case that the DNA database program constituted such a special need, the *Roe* court looked to a 1987 Supreme Court case, *Griffin v. Wisconsin.*[10] In *Griffin*, the Supreme Court reasoned that a state's operation of a probation system looked sufficiently like its operation of schools, government offices and prisons—other situations in which the special needs reasoning had been used to justify generalized regulatory searches. In discussing this precedent, the *Roe* court singled out one piece of the Supreme Court's reasoning in particular, namely empirical "research that indicated that more intensive supervision of probationers reduced recidivism" (*Roe*, 79). The *Roe* majority seized upon the idea that the aim of reducing recidivism helped the Wisconsin statute qualify for a "special needs" exemption.

After its discussion of *Griffin*, the *Roe* court concluded that the statute indeed passed constitutional muster, stating that "a reasonable interpretation of the 'special needs' doctrine supports the constitutionality of the DNA statute" (79). The court proceeded with a balancing test of the interests implicated by the statute. Motivated by the concerns of Judge Murnaghan in his *Jones* minority opinion, the *Roe* majority pointed first and foremost to social science suggesting that the database would be especially efficacious with respect to the included populations: "In defense of the statute, defendants cite studies indicating a high rate of recidivism among sexual offenders. Moreover, DNA evidence is particularly useful in investigating sexual offenses and identifying the perpetrators because of the nature of the evidence left at the scenes of these crimes and the demonstrated reliability of DNA evidence" (79). Balanced against these interests, and a general interest in deterring crime, the court assessed "an intrusion that the Supreme Court has characterized as minimal," namely the "drawing for blood for testing."[11] Notably, the *Roe* court diverged from the *Jones* analysis by specifically considering statutory safeguards enacted "to ensure that the intrusion is minimal," including regulations on the handling and analysis of blood, restriction of access to and confidentiality of the database, and provision for expungement of the profile from the database after reversal or dismissal of a conviction (80). Nevertheless, the *Roe* court concluded that the special needs balancing weighed in favor of constitutionality, a result that many other courts have followed.

"Individualized Suspicion"

Just as the majority of federal and state courts were settling into a pattern of affirming the constitutionality of forensic DNA database statutes, the Ninth Circuit dropped a bombshell in October 2003. In *United States v. Kincade (Kincade I)*, a three-judge panel of the Ninth Circuit ruled that the DNA Analysis Backlog Elimination Act (DABEA), the federal DNA statute enacted in 2000, was unconstitutional under the Fourth Amendment.[12] The decision caused a firestorm in the law enforcement community and among forensic scientists. Importantly, this decision was vacated soon after and the case was reheard *en banc* by the full Ninth Circuit (*Kincade II*). A plurality opinion reversed the first decision, declaring that the DABEA passed constitutional muster after all.[13] But in his majority opinion in *Kincade I*, and in his dissent in *Kincade II*, Judge Reinhardt strongly articulated a third possible doctrinal treatment for forensic DNA databases.

In the *Kincade* cases, a parolee appealed from a sentence imposed by the district court for his refusal to comply with a compulsory blood extraction pursuant to the DABEA. The three-judge panel split 2–1, with Judge Reinhardt, known to be a strong supporter of civil liberties, writing on behalf of Judge Paez for the majority and Judge O'Scannlain writing a strong dissent. The main holdings of the case were twofold: (1) "forced blood extractions from parolees pursuant to the database act required individualized suspicion" and (2) the "special needs doctrine" did not apply.[14]

The legal reasoning in *Kincade I* began, as all the others, by asserting that searches requiring mandatory blood extraction constitute an intrusion into an area, the body, normally protected by the Fourth Amendment. But Judge Reinhardt went further in characterizing the social significance of this intrusion, stating, "In virtually every culture in the world, human blood possesses great symbolic power, and its spillage—whether in a drop or in a torrent—has carried enormous cultural significance. Throughout history, we have waged war, organized societies and religions, and created myths based upon the substance."[15] That opening remark laid the groundwork for a centerpiece of Reinhardt's argument, namely that the analogy between fingerprinting and blood draws is a "false" one. The differences between fingerprinting, which requires the "examination or recording of physical attributes that are generally exposed to public view," and DNA profiling, requiring a "forced intrusion into an individual's body," are constitutionally significant (*Kincade I*, 1100).

According to Judge Reinhardt in *Kincade I*, the nature of the intrusion triggered a typical Fourth Amendment analysis. But rather than jumping to a formal analysis of whether DABEA falls within a legitimate exception to a warrant requirement, Judge Reinhardt conducted a balancing test. Unlike nearly every other court, this opinion characterized the intrusion as "substantial," emphasizing bodily integrity as a "cherished value of society," and that parolees' privacy rights in their bodies, though diminished, were not extinguished (*Kincade I*, 1102). But neither were they absolute. Reinhardt stated that the purpose of obtaining samples

is to further "the overwhelming public interest in creating a comprehensive nationwide DNA bank that will improve the accuracy of criminal prosecutions" for generations to come. It is undoubtedly true that, were we to maintain DNA files on all persons living in this country, we would even more effectively further the public interest in having efficient and orderly criminal prosecutions, just as we would were we to sacrifice all of our interests in privacy and personal liberty. (1103)

Reinhardt thus concluded that the government's interest did not outweigh Kincade's privacy interest "in his body." But he went one step further, saying that all of the considerations in the balance of interests must be weighed "in light of the fact . . . under controlling Supreme Court authority we are not free to approve suspicionless searches conducted for law enforcement purposes" (1103). Though the results of the balancing test might affect the "degree of suspicion or cause required to conduct such searches, it could not serve to eliminate the requirement of individualized suspicion entirely" (1103). For Reinhardt, the need to find individualized suspicion remained important, resisting the logic of the previous DNA database cases that swept this requirement away in favor of generalizing suspicion across all classes of criminal subjects.

A significant portion of Reinhardt's opinion in *Kincade I* is devoted to rebutting the availability of the "special needs" doctrine, a conclusion that few legal scholars disagree with (Maclin 2005; Carnahan 2004, Kaye 2006). In *Kincade I*, the government's own characterization of the DNA statute's "primary purposes" were "to help law enforcement solve unresolved and future cases," and "to increase accuracy in the criminal justice system." Reinhardt used these statements to conclude that the government's goals were nothing more than the normal need for law enforcement, therefore putting the special needs exception out of reach.

As mentioned previously, after the Ninth Circuit declared DABEA unconstitutional in *Kincade I*, the decision was vacated and the case was reheard *en banc*. In a plurality opinion, the *en banc* court upheld the constitutionality of the DABEA. As in *Jones*, the Ninth Circuit applied a "totality of the circumstances" test to find the search "reasonable" given the substantially diminished expectation of privacy by the parolee, the minimally intrusive nature of blood sampling, and the important social interest "furthered by the collection of DNA" (*Kincade II*, 839). The court found that a special needs analysis was not required in determining the constitutionality of the statute. Judge Reinhardt reiterated his "individualized suspicion" doctrine in a scathing dissent, analyzed in more detail in the following section.

Analysis

After *Kincade II*, the circuit courts are no longer split on the question of whether the collection of DNA from those enmeshed in the criminal justice system violates their Fourth Amendment rights. The Supreme Court recently refused an opportunity to clarify the law further when Kincade appealed the Ninth Circuit's *en banc* ruling. Nevertheless, doctrines diverge in interesting ways. Close attention to the *dicta* in these cases (that is, their nonbinding language) suggests a close interaction between the technological and legal imagination.

First, the judges in these cases conceive of technology and its potential impact on civil liberties quite differently from one another. They engage in a form of technological risk assessment, in which short- and long-term hazards to those within the criminal justice system, and to society as a whole, need to be imagined and balanced against potential benefits. These assessments, in turn, depend upon competing conceptions of the technology in question, variously described as a fingerprint, a type of information, and a large state-operated surveillance network. These doctrinal choices and risk assessments also depend on prior conceptions of how, as a general matter, the Fourth Amendment responds to new technologies. In particular, there is a stark split between those judges who imagine the continuous development of law enforcement technologies as benign, and those who see a malignant trend that slowly erodes Fourth Amendment protections. These views take different positions on the trade-off between individual freedom and collective security.

Second, these technological imaginaries help condition and construct the legal identities of criminal kinds. Thus, the "reasonableness" approach simultaneously constructs a newly legalized forensic technology and a broad and undifferentiated set of future "usual suspects," from convicts to arrestees. By contrast, the "special needs" approach constructs a narrower set of subjects, the recidivists, who justifiably suffer the brunt of the extra privacy intrusion. The "individualized suspicion" approach resists reifying these criminal kinds. Conditioned by a pessimistic paradigm of technological authoritarianism, this last position resists normalizing the technology as an unproblematic intrusion on those who have already, to some degree, forfeited their freedom.

"Totality of the Circumstances"

In *Jones*, Judge Niemeyer characterized the new forensic DNA technology as a high-tech fingerprint: no more threatening to privacy than the normal booking procedure but much more powerful and precise than existing fingerprinting. Niemeyer emphasized the way in which nucleotides "are arranged differently for every individual except for identical twins" (*Jones*, 303). This set up his statement that "improved scientific technology has prompted efforts to use the individuality of a person's DNA in the context of criminal law enforcement," and he proceeded to describe the "DNAPrint": the digital and/or visual representation of a set of thirteen short tandem repeats (STRs) on the human genome analyzed for an entry onto any forensic DNA database (304). Today, "DNA Profile" is the dominant name.

Fingerprinting is the key analogy in the opinion. As we have already seen, the use of fingerprints ends up providing the foundation for his constitutional argument. Because the Virginia statute authorizes a blood draw, there is a search involved in using this technology, but because it is merely a "limited search for the purpose of ascertaining and recording the identity of a person who is lawfully confined to prison," it is functionally analogous to the constitutionally accepted practice of fingerprinting, and therefore deemed minimally invasive (*Jones*, 306).

A logic of equivalence drives Niemeyer's tacit risk assessment of the new technology. For the judge, the DNA databases do not differ significantly from forensic fingerprinting, which also involves the indefinite storage of bodily information, searchable at a click. But Niemeyer neglects to consider the ways in which fingerprints, though also undergoing the material to informational shift, remain a different kind of thing from DNA. Prints are not bodily material, they are not potentially health-related, and they do not implicate genetic relatives. For these reasons, DNA can be considered special, a finding that was reinforced (albeit in a very different context) by a widely discussed gene patenting case in 2010.[16] Further, the judge's analysis of the degree of intrusiveness of the search only considers the blood draw, not the continuing technological surveillance, the banking of DNA samples indefinitely, and the frequent searches of biometric information.

Niemeyer's technological optimism carries the day. Whereas the new technique poses few new risks to the incarcerated, and none to greater society, it does bring significant new benefits. The power of this "dramatic new tool," increased precision, helps justify what has already been sanctioned in the fingerprint context: "The government justification for this form of identification, therefore, relies on no argument different in kind from that traditionally advanced for taking fingerprints and photographs, *but with additional force because of the potentially greater precision of DNA sampling and matching records*" (307; emphasis added).

The new technology becomes simply a better case of an adequately controlled technology and well-known practice. Just as genetically modified crops were normalized by the Food and Drug Administration through an across-the-board determination of "substantial equivalence," so too a judgment of equivalence operates to normalize the DNA database within legal and cultural logics. Having read the DNA technology as equivalent to the fingerprint, Niemeyer deemed the objective of the Virginia statute "significant" and "the privacy intrusion limited" (*Jones*, 308).

This logic of controlled equivalence has been evident throughout the expansion of forensic databases, from including only felons convicted of violent crimes to those convicted of lesser offenses, parolees, and now arrestees. An unstated logical implication of this rule, however, is that arrestees of particular crimes—not just "all convicted felons"—would become fair game for DNA profiling and banking because they are sufficiently suspicious to satisfy a probable cause standard for future database searches.

Such a logical implication could not have escaped the *Jones* majority. It is too obvious. And indeed the case has provided the doctrinal structure and rationale for a number of courts to uphold arrestee inclusion statutes, including the Virginia Supreme Court in 2007 and a California District Court in 2009.[17] In effect, the risk management strategy is to limit the threat of civil liberty deprivation to convicts, and at its broadest, arrestees, as this population is deemed, in turn, to pose unacceptable risks to society. Such a rule carries clear potential for increasing existing inequalities in the criminal justice system. Because blacks and other minorities are more often targeted for pretextual arrests and police profiling, they will bear a disproportionate amount of intrusion (Kaye and Smith 2003, n153). Most often, blacks and the poor will be in the database, and whites and the privileged out.

"Special Needs"

The special needs opinions discussed previously provide a contrast with the technological imaginary of the *Jones* majority. In the *Jones* dissent, Judge Murnaghan expressed skepticism toward the majority's blanket assumptions about the efficacy of the new technology. With the danger of database expansion in mind, these special needs opinions carve out a narrow exception to the individualized suspicion rule for felons who have perpetrated crimes with high recidivism rates. Rather than identifying the prisoner/nonprisoner boundary as relevant for Fourth Amendment protection, the "recidivist" rule posits subcategories of felons discernable through empirical findings and scientifically grounded classification. The general presumption against inclusion is maintained, creating a more precautionary approach with respect to the general erosion of civil liberties.

Judge Murnaghan exhibited a general wariness toward what he perceived as a dangerous trend in the erosion of civil liberties in the face of creeping police power. He was concerned that under the majority's logic, the "disturbing restriction of the Fourth Amendment protections afforded to the nation's prisoners" could easily be extended to all citizens (*Jones*, 313). Murnaghan's explicit invocation of "citizens in a free society still clinging to disappearing Fourth Amendment protections" painted a picture of liberties in grave danger. The judge confessed to "a deep, disturbing and overriding concern that, without proper and compelling justification, the Commonwealth may be successful in taking significant strides towards the establishment of a future police state, in which broad and vague concerns for administrative efficiency will serve to support substantial intrusions into the privacy of citizens" (315).

To place brakes on this slippery slope, he looked to the empirical data before the court, which included two reports on prisoner recidivism,[18] to establish a constitutionally relevant distinction between violent and nonviolent felons for purposes of database inclusion. Such reasoning had no precedent in the Fourth Circuit, but it proved persuasive to the *Roe* court and became law in the Second Circuit.

Judge Pooler's opinion for a unanimous three-judge panel in *Roe* grappled with further details of the statute with regard to the handling of physical samples and the continued use of DNA profiles. This analysis explicitly acknowledged that more is at risk than the intrusion into criminal bodies and brought ongoing possession of samples and use of information into the frame of judicial risk assessment. However, as noted previously, the unanimous *Roe* panel was satisfied with the statute's mitigating safeguards (e.g., securing the confidentiality of the results, and providing for expungement of profiles upon reversal). These safeguards help contain the risk of slippage toward a police state, but they are complemented by another critical management strategy: a legal boundary between the recidivist and nonrecidivist classes of criminals. This boundary establishes a firewall against the possible expansion of surveillance from the criminal classes to the rest of society, from "them" to "us." This position embodies a general distaste for blanket suspicionless searches, but creates a limited exception where there is, in essence, "statistical probable cause" to believe that particular groups (i.e., recidivists) will commit future crimes.

Embedded within this risk management strategy, we see a particular characterization of the Fourth Amendment-technology interface, that is, as a system that is controllable through legal rules based on good social science. The view relies on a prediction model to justify inclusion: because recidivists are likely to commit more crimes, inclusion in the database may deter this class of individuals, or help us take them off the street. The fact that this form of statistical probable cause needs to be established through scientific inquiry affords society its necessary protection against universal inclusion and a "future police state."

"Individualized Suspicion"

In *Kincade I*, Judge Reinhardt focused on the blood draw as a clear privacy intrusion requiring some degree of individualized suspicion in lieu of a warrant. In his *Kincade II* dissent, he focused more attention on the nature of the technology itself. For Reinhardt, forensic DNA is not simply a booking tool like a fingerprint, but a large centralized technological system for rounding up the usual suspects and a repository of potentially sensitive information. In both opinions, Reinhardt considered the case against a backdrop of judicial suspicion toward new surveillance technologies, a tradition that recognizes a slippery slope toward the erosion of privacy. The individualized suspicion approach draws a firm line in the sand against the construction of forensic DNA databases, refusing to corral felons or prisoners en masse for inclusion.

In characterizing CODIS, Reinhardt described what Langdon Winner would call an artifact with politics (Winner 1986), a technology that necessitates a particular form of political power: in this case a strongly authoritarian police state that can bully vulnerable elements of the public. The majority's decision in *Kincade II*, Reinhardt argued, "encourages the very centralization of government authority that has repeatedly resulted in the sacrifice of our liberties in the name of law enforcement" (*Kincade II*, 843–844). Accordingly, he drew comparisons to the abuses of J. Edgar

Hoover in monitoring civil rights leaders in the 1960s, and the country's use of central databases to "round up" Japanese-Americans and Communists during the 1940s and 1950s (843). Furthermore, in contrast to the judges in *Jones* and *Roe*, Reinhardt dwelt on the nature of the information contained in the DNA profiles and the retention of samples under the statutes. He emphasized the sensitivity of biometric information and also the fact that, "as technology evolves," the maintenance of DNA "will permit a myriad of other known and unknown uses of the samples (870).

Reinhardt projects these images of forensic DNA profiling onto an existing tradition of constitutional concern with new surveillance technologies. In its first sentence, Judge Reinhardt's majority opinion in *Kincade I* emphasizes the special hazards posed by technological developments to Fourth Amendment protections: "Each leap forward in forensic science promises ever more efficient and swift resolution of criminal investigations. At the same time, technological advances frequently raise new constitutional concerns and threaten our basic liberties" (*Kincade I*, 1096).

To support this statement, Reinhardt cites the 2001 Supreme Court case of *Kyllo v. United States*.[19] Writing for the 5–4 majority in that case, Justice Scalia maintained "it would be foolish to contend that the degree of privacy secured to citizens by the Fourth Amendment has been entirely unaffected by the advance of technology" (*Kyllo*, 33–34). In that case, law enforcement agents suspected that marijuana was being grown in a home belonging to Danny Kyllo. Without a warrant, agents used infrared-detecting thermal imagers to scan Kyllo's apartment for heat emissions typical of lamps needed to grow marijuana indoors. Justice Scalia held that the surveillance constituted a Fourth Amendment search, and was therefore presumptively unreasonable without a warrant. He cited the 1961 case *Silverman v. United States*, in which the court notes the "frightening paraphernalia which the vaunted marvels of an electronic age may visit upon human society."[20] Reinhardt in turn invoked this pattern of technological skepticism in *Kincade I*, sharing with Judge Murnaghan a fear of the loss of a free society.

It is Judge Kozinski in *Kincade II*, not Reinhardt, who in dissent imagined the slippery slope in the starkest terms. He emphasized the power of a tempting new technology to lure society onto the slippery slope by its very efficiency. Born in Communist Romania of parents who were Holocaust survivors, Kozinski criticized what he called the plurality's "exuberant faith in the positive power of technology" (*Kincade II*, 872), and discussed the attitude-skewing effects of the database's crime-solving power: "Later, when further expansions of CODIS are proposed, information from the database will have been credited with solving hundreds or thousands of crimes, and we will have become inured to the idea that the government is entitled to hold large databases of DNA fingerprints" (873).

He called the plurality's opinion an "engraved invitation" to future expansions of the database, and predicted that each step may not seem like much until "the fishbowl will look like home" (*Kincade II*, 874). Accordingly, he declared that "the time to put the cork back in the brass bottle is now—before the genie escapes" (875). For Kozinski, technological advance, at least in surveillance, is a special source of risk to Fourth Amendment protections. Under the "reasonable expectations" standard in *Katz*, the scope of privacy protection depends on the subjective expectations of individuals, meaning that the government could eliminate privacy rights by accustoming the public to heightened levels of technological surveillance.

Reinhardt agreed with Kozinski that DNA profile technology exerts a sort of malignant gravitational pull toward the bottom of the slippery slope, noting with concern that some state statutes are already providing for the inclusion of arrestees and those convicted of misdemeanors. If "totality of the circumstances" should become the general rule, "we all have reason to fear that the nightmarish worlds depicted in films such as *Minority Report* and *Gattaca* will become realities" (*Kincade II*, 851). At the bottom of this slide lies the possibility that "the database could be used to repress dissent or, quite literally, to eliminate political opposition" (847).

Accordingly, Reinhardt and the other dissenters in *Kincade* cleaved to a bright line of individualized suspicion. All other doctrinal dispensations, they argue, "dismantle the structural protections that lie at the core of the Fourth Amendment" and simply ask us "to trust those in power" (845). Reinhardt identified a strong precautionary logic against the aggrandizement of power in the structural checks and balances of the Constitution. Whereas other judges managed the Fourth Amendment–technology interface by creating new kinds of suspect classes insulated from the innocent public, Reinhardt's approach would mean that—with respect to bodily intrusions for DNA databanks—felons, arrestees, and other classes would not be systematically downgraded. Instead, the individualized suspicion standard would continue to apply, maintaining the individual-state boundary as the crucial one for legal analysis. In other words, the arrestee is still one of "us" where civil liberties are at stake, and so too is the convict.

Conclusion

In this chapter, I have examined the play of imagination as different judges weigh the constitutionality of a new technology. We have seen how different technological imaginaries form part of the interpretive framework of judging, conditioning and shaping doctrinal choice and legal reasoning. Some see a lurking dystopia in the new technological order; others see prospects for a safer society. These imaginaries entail not only visions, theories, and a priori characterizations of technological objects, but also models of how those objects interact with the social technology known as the law. Some judges recognize technologies to be problematic objects for legal analysis precisely because they develop and mutate; others consider them to be stable and predictable. These positions affect judicial calculations of the risks and benefits of selecting particular legal rules. Risk management strategies must address how new technological orders will engage the constitutional architecture in the long term. Thus, we see judges trying to manage the interoperability of two intersecting technological systems—one material, one normative.

As these Fourth Amendment cases demonstrate, legal adjudication can be a process and medium through which social and technological orders develop together and receive new articulation. In other words, it is site of coproduction, and an especially important one. These judicial imaginaries have stark consequences for different classes of citizens, who become unwitting players in the dynamics of bioconstitutionalism. The stakes are high. Even as these judicial imaginaries help bring new technological orders into full operation, they produce new kinds of subjects whose rights are reframed without their direct participation.

Notes

1. *Katz v. United States*, 389 U.S. 347, 362 (1967) (Harlan, J., concurring).

2. *California v. Acevedo*, 500 U.S. 565, 580 (1991).

3. *Mincey v. Arizona*, 437 U.S. 385, 390 (1978), quoting *Katz*, 389 U.S. at 357 (footnotes omitted).

4. *Robbins v. California*, 453 U.S. 420, 438 (1981) (Rehnquist, J., dissenting).

5. *En banc*, French for "in the bench," signifies a decision by the full court of all the appellate judges on the court. This process is often invoked when there is a particularly significant issue at stake or when requested by a party to the case and agreed to by the court.

6. *Jones v. Murray*, 962 F.2d 302 (4th Cir. 1992).

7. See, for example, *Shaffer v. Saffle*, 148 F.3d 1180, 1181 (10th Cir. 1998); *Boling v. Romer*, 101 F.3d 1336, 1340 (10th Cir. 1996); *Schlicher v. Peters*, 103 F.3d 940, 943 (10th Cir. 1996); *Rise v. Oregon*, 59 F.3d 1556, 1560–1562 (9th Cir. 1995); *Kruger v. Erickson*, 875 F. Supp. 583, 588–589 (D.Minn. 1995); *Sanders v. Coman*, 864 F.Supp. 496, 499 (E.D.N.C. 1994); *Ryncarz v. Eikenberry*, 824 F. Supp. 1493, 1498–1499 (E.D.Wash. 1993).

8. *Roe v. Marcotte*, 193 F.3d 72, 81 (2nd Cir. 1999).

9. *Roe v. Marcotte* at 77, quoting and citing *Skinner v. Railway Labor Execs. Association*, 489 U.S. 602, 624 (1989).

10. *Griffin v. Wisconsin*, 483 U.S. 868 (1987).

11. *Griffin v. Wisconsin* at 79, citing *Skinner*, 489 U.S. at 625.

12. *U.S. v. Kincade* (*Kincade I*), 345 F.3d 1095 (9th Cir. 2003).

13. *U.S. v. Kincade* (*Kincade II*), 379 F.3d 813 (9th Cir. 2004) (*en banc*).

14. *Kincade I*, 345 F.3d at 1095.

15. *Kincade I* at 1099–1100, citing Nelkin 1999, 110.

16. *Association for Molecular Pathology v. United States Patent and Trademark Office*, No. 09 Civ. 4515 (S.D.N.Y., Mar. 29, 2010).

17. *Anderson v. Commonwealth*, 274 Va. 469 (Va. 2007) and *U.S. v. Pool*, 645 F. Supp. 2d 903 (E.D. Cal. 2009). In contrast, a Minnesota court of appeal has held that arrestee inclusion is unconstitutional for lack of probable cause. *In re Welfare of C.T.L.*, 722 N.W. 2d 484 (Minn. Ct. App. 2006).

18. The reports cited in the opinion were Beck and Shipley 1989 and Virginia Division of Justice and Crime Prevention 1989.

19. *Kyllo v. United States*, 533 U.S. 27 (2001).

20. *Silverman v. United States*, 365 U.S. 505, 509 (1961).

References

Aronson, Jay. 2006. *The Introduction, Contestation, and Regulation of Forensic DNA Analysis in the American Legal System (1984–1994)*. Minneapolis, Minn.: University of Minnesota Press.

Beck, Allen J., and Bernard E. Shipley. 1989. "Recidivism of Prisoners Released in 1983." Bureau of Justice Statistics Special Report: 1–13.

Biancamano, John D. 2009. Arresting DNA: The Evolving Nature of DNA Collection Statutes and Their Fourth Amendment Justifications. *Ohio State Law Journal* 70:613–660.

Bieber, Frederick. 2004. Science and Technology of Forensic DNA Profiling: Current Use and Future Directions. In *DNA and the Criminal Justice System*, ed. David Lazer, 23–62. Cambridge, Mass.: MIT Press.

Bradley, Craig M. 1993. The Court's "Two Model" Approach to the Fourth Amendment: Carpe Diem! *Journal of Criminal Law & Criminology* 84:429–461.

Callon, Michael. 1987. Society in the Making: The Study of Technology as a Tool for Sociological Analysis. In *The Social Construction of Technological Systems*, ed. Wiebe E. Bijker, Thomas P. Hughes, and Trevor Pinch, 83–103. Cambridge, Mass.: MIT Press.

Carnahan, S. J. 2004. The Supreme Court's Primary Purpose Test: A Roadblock to the National Law Enforcement Database. *Nebraska Law Review* 83:1–37.

Hughes, Thomas. 2004. *Human-Built World: How to Think About Technology and Culture*. Chicago, Ill.: University of Chicago Press.

Jasanoff, Sheila. 2008. Making Order: Law and Science in Action. In *New Handbook of Science and Technology Studies*, ed. Ed Hackett et al., 761–786. Cambridge, Mass.: MIT Press.

Jasanoff, Sheila, and Sang-Hyun Kim. 2009. Containing the Atom: Sociotechnical Imaginaries and Nuclear Power in the United States and South Korea. *Minerva* 47 (2): 119–146.

Kaye, David H. 2006. Who Needs Special Needs? On the Constitutionality of Collecting DNA and Other Biometric Data from Arrestees. *Journal of Law, Medicine & Ethics* 34:188–198.

Kaye, David H., and Michael E. Smith. 2003. DNA Identification Databases: Legality, Legitimacy, and the Case for Population-Wide Coverage. *Wisconsin Law Review* 2003:413–459.

Lynch, Michael, Simon A. Cole, Ruth McNally, and Kathleen Jordan. 2008. *Truth Machine: The Contentious History of DNA Fingerprinting*. Chicago, Ill.: Chicago University Press.

Maclin, Tracey. 1994. When the Cure for the Fourth Amendment Is Worse than the Disease. *Southern California Law Review* 68:1–72.

Maclin, Tracey. 2005. Is Obtaining an Arrestee's DNA a Valid Special Needs Search Under the Fourth Amendment? What Should (And Will) the Supreme Court Do? *Journal of Law, Medicine & Ethics* 33 (1): 102–124.

Murphy, Erin. 2008. Paradigms of Restraint. *Duke Law Journal* 57:1321–1411.

Nakashima, Ellen, and Spencer Hsu. "U.S to Expand Collection of Crime Suspects' DNA—Policy Adds People Arrested But Not Convicted." *Washington Post*, April 17, 2008, A1.

Nelkin, Dorothy. 1999. Cultural Perspectives on Blood. In *Blood Feuds: AIDS, Blood, and the Politics of Medical Disaster*, ed. Eric A. Feldman and Ronald Bayer, 273–292. New York: Oxford University Press.

Power, Robert C. 1989. Technology and the Fourth Amendment: A Proposed Formulation for Visual Searches. *Journal of Criminal Law & Criminology* 80:1.

Virginia Division of Justice and Crime Prevention. 1989. *Violent Crime in Virginia*. Richmond: Commonwealth of Virginia, Department of Criminal Justice Statistics.

Virginia Joint Subcommittee Studying Creation of a DNA Test Data Exchange. 1990. Report of the Joint Subcommittee Studying Creation of a DNA Test Data Exchange to the Governor and the General Assembly of Virginia. Richmond: Commonwealth of Virginia.

Winner, Langdon. 1986. Do Artifacts Have Politics? In *The Whale and the Reactor: A Search for Limits in an Age of High Technology*, ed. Langdon Winner, 19–39. Chicago, Ill.: University of Chicago Press.

Wynne, Brian. 1988. Unruly Technology: Practical Rules, Impractical Discourses and Public Understanding. *Social Studies of Science* 18 (1): 147–167.

8

Risks and Rights in Xenotransplantation

Mariachiara Tallacchini

I don't think we are going to really feel comfortable . . . until we have done 1,000 patients or maybe even 10,000 patients, all the animal studies aside. We really just need to do patients.
—D. Auchincloss (FDA 1997)

Well, that is a tough one when you are saying, you know, the patient is going to die unless we do a xenotransplant. They will probably truly die with one.
—H. Y. Vanderpool (FDA 1997)

Started as a form of experimental surgery and advertised since the 1960s as a solution to the shortage of human organs, xenotransplantation (XT), the transplant of cells, tissues, or organs between different species, moved toward clinical application during the 1990s, when developments in genetic engineering made transgenic animals available for this purpose. The major technical obstacle to performing transplants from one species to another, the hyperacute rejection (HAR) of the transplanted organ or tissue—also existing, but in a weaker form, in human-to-human transplantation—was at least partially overcome when pig tissues and organs (the species of choice for this technology) became more compatible with humans through genetic engineering.

But the ups and downs of xenotransplants as an acceptable therapeutic approach have depended more on factors other than mere feasibility. The most controversial and still unresolved point concerns the ethical use of animals, both pigs and nonhuman primates used as recipients in preclinical trials (Nuffield Council on Bioethics 1996). Numerous national and international documents addressing the issue have invariably agreed on the acceptability of using pigs and for increasing restrictions in experimenting with nonhuman primates.

Besides the scientific and ethical barriers relating to animal use, XT raises concerns about both individual and collective rights, especially in the conduct of clinical trials. Because of the potential risk of infections transmitted from the animal source to the human recipient, XT breaks the common rules of individual informed consent. On the one hand, the lack of adequate information makes informed consent inconceivable and lays the basis for additional constraints to which patients have to consent; on the other hand, threats to collective safety from the possibility that novel or known infectious agents (so-called xenogeneic infections) (Chapman et al. 1995) may spread to the population at large not only require new ways of thinking about individual and collective rights, but also represent a challenge for bioethics in individually oriented liberal democratic societies.

At the beginning of the twenty-first century, as conditions for the delicate passage from preclinical to clinical trials seemed near, different regulatory frameworks were developed to normalize XT, primarily by mitigating the potential risks of infections while protecting involved subjects. This paper explores four legal and policy approaches to XT framed in that period. These are the U.S. (2001), the European (Council of Europe and European Union) (2003), the Canadian (2002), and the Australian regulations (2004) (the third and fourth as variants of the same model). Each uses a different strategy to build legitimacy for a technology whose implementation requires well-established rights to be reframed (Jasanoff 2004; Jasanoff 2005; Jasanoff, chapter 1, this volume).

Each of these legal approaches had to adjust both science and norms to create an acceptable context for XT: although certain legal constraints were justified on technical grounds, specific scientific statements were adopted not because of their factual reliability but rather for their expected legal implications. Each set of adjustments has given rise to a particular scheme of cognitive and social order, new roles and identities for citizens, and new social relationships. Collectively, these models have opened up broader and more complex scenarios of legitimate government for controversial biomedical technologies.

At stake in the U.S. regulatory model is the maintenance of a coherent, contractual, liberal vision of society. In the European Union (EU) and Council of Europe (COE) model, the paternalistic and vertical construction of Europe as a fully political entity, responsible for European citizens, represented the main challenge (see also Dratwa, chapter 12, this volume). In the cases of Canada and Australia an incomplete attempt was made to extend the concept of citizenship and so to integrate the realities of state and society.

Historically, XT was often associated with the chimera (a mythological figure made of different animal parts). However, XT has constantly changed its discursive identity. From relying on organ shortage as its main rationale and being primarily framed as a surgical technique (Cooper and Lanza 2000; McLean and Williamson 2005), XT has been reconstructed as "advanced cell therapy" and in Europe is now classified as pharmaceutical product (EMA 2009). These new conditions are going to redefine the normative landscape for XT, which now may become a widely applied therapy, potentially affecting large populations in a globalized regulatory context (WHO 2009). Though bioethicists are now facing the need to invent an ethics for pandemics—so-called *pandethics* (Selgelid 2009)—capable of harmonizing individual and collective rights, xenotransplantation is today in the unique position of having triggered relevant schemes for public health protection. In this respect, XT not only represents a remarkable case for comparative analysis, but also offers rich materials for future bioconstitutional reframing.

More than a chimera, XT is a chameleon: a technology that changes its aspect and repeatedly emerges and reemerges while raising new problems of rights and responsibilities.

Constructing a Socially Acceptable Science

A culture of secrecy accompanied the first experiments in transplant surgery. Thomas Starzl only lately revealed that he performed the first liver transplant in 1963, not publicly disclosed until much later (ABC 1995–1996). Through the years, this culture of secrecy has been replaced, as transparency and openness in the governance of technological innovation have risen in political significance. The attempt to create a credible science and a favorable cultural environment for XT became more apparent as this biomedical technology entered public debate.

The modern history of XT began in the 1960s. The first partially successful vascularized organ xenotransplants were performed in the early 1960s by Keith Reemtsma, who transplanted chimpanzee kidneys to six humans at New York's Columbia Presbyterian Medical Center. The first patient survived sixty-three days. In 1964, Starzl transplanted baboon kidneys to six humans (Reemtsma et al. 1964). Between 1966 and 1973, he also transplanted chimpanzee livers into three children; the longest survival was fourteen days (Starzl et al. 1964).

Public awareness of XT is linked to the 1984 Baby Fae case (Bailey et al. 1985; Annas 1985; Kushner and Belliotti 1985). Baby Fae was born with severe heart failure—a hypoplastic left heart syndrome. She received a baboon heart transplant and survived twenty-one days. Newspaper headlines initially presented the experiment with enthusiasm: "Baby Fae stuns the world" (Wallis 1984a, 70). The child's death was initially described as an intentionally accepted risk: "Baby Fae loses her battle" (Wallis 1984b, 88). Subsequent reaction, however, both in academic and public spheres, brought a temporary halt to research or at least to the "candid" enthusiasm attached to it (Bailey 2005). In order to control the progressive narrative on XT, scientists endorsed a hagiographic remaking of the past, suggesting that "promises are certain and perils uncertain" (Weiss 2000), emphasizing potential benefits ("pig xenografts, currently grounded, might eventually fly!") (Dorling 2002) and describing pioneering surgeons as heroes (Najarian 2003).

Though other transplants took place in the early 1990s, later trials concentrated on tissues and cells more than organs because of safety issues and little evidence of efficacy. The major factor leading XT to regain attention (especially that of the market) was the development of transgenic animals and their potential to overcome HAR. In the mid-1990s the switch from the researcher-driven to the industry-driven phase of XT experimentation changed the image of the technique. Research in the earlier phase was performed by prominent surgeons transplanting whole organs; in the later phase, research has been increasingly undertaken by biotech and pharmaceutical companies, focusing primarily on cell and tissue grafting in clinical experimentation. Currently, the most promising field for XT is the implant of swine pancreatic islets as a therapy against diabetes (Groth and Korsgren 2008).

In 1997, after studies reported that human cells had been infected in vitro by pig retroviruses—that is, porcine endogenous retroviruses (PERVs) that are permanently integrated into the pig genome (Patience et al. 1997)—XT began attracting critical attention from international organizations. In 1998, a call for a moratorium by some leading XT scientists (Bach et al. 1998; Bach and Fineberg 1998) fractured the scientific community between supporters and critics (Fishman 1998; Paradis et al. 1999). At this point, regulation took the lead, becoming a major factor in the further development of XT.

Normalization: Regulatory Frameworks for XT

Each regulatory model discussed here presents characteristic features that, in effect, normalize XT within a larger constitutional order. The U.S. guideline reflects the individualistic model of rights and responsibilities stemming from the field of human subjects research, a contractual vision strongly influenced by the history of experimentation (Vanderpool 2007). The Council of Europe, which worked out its position in collaboration with the EU Commission, relies on a public health argument to delete individual liberties, showing how European institutions are using techno-science to construct and implement their ideas of citizenship while constructing European society itself. The Canadian and the Australian approaches apparently accept the radical challenge posed by scientific uncertainty in an active democratic order. However, though interestingly applying new deliberative procedures and reflecting on their strengths and weaknesses, they repropose communitarian approaches that limit individual freedom through internal controls.

Because of its potential to spread epidemics to the population at large, the history of XT in all these legal systems is deeply connected to the rise of the precautionary principle (PP), originally articulated so that "lack of full scientific certainty" would not be used as a reason to block protective regulation (on the drafting of the PP in Europe, see Dratwa, chapter 12, this volume). In the debate following the release of the 1996 U.S. draft guideline (FDA, DHHS, and PHS 1996), the Nuffield Council on Bioethics (1996) proposed to apply the precautionary principle to XT. Though that position never officially took hold, some prominent U.S. scientists advocated a moratorium on research (Bach et al. 1998), other international documents expressed concern about XT, and the Council of Europe in 1999 also called for a moratorium.

Very quickly, however, the idea of a moratorium was dismissed. In the United States, precaution was replaced by cautious management of risks. In the European context, precaution toward XT was reinterpreted as precaution against transplanted patients, who represented the real threat. Canada and Australia explored a different meaning of precaution by advocating public participation to fight the political risks of uncertainty.

The U.S. Liberal Model

On January 19, 2001, the Public Health Service (2001) released the *Guideline on Infectious Disease Issues in Xenotransplantation*, five years after the first draft was proposed. The document defined the practices required in clinical trials in order to prevent and control potential infectious diseases associated with XT. The technical nature of the guideline implied that the document was not meant to deal with rights and responsibilities. Consideration of the full range of normative issues raised by the technology was the task of the Secretary's Advisory Committee on XT (SACX), a response to the deep social concerns rising around XT.

The lengthy comment period in the PHS guideline and the formation of SACX signaled an opening up to the social aspects of XT (DHHS-SACX 2004). At the same time, the attempt to separate facts and values in XT regulation reiterated a well-established character of U.S. science policy, namely the tendency to produce firm boundaries between the technical and legal-political aspects of decision making (Jasanoff 2005).

Two parties dominate the U.S. regulatory approach and are considered responsible for any problem: the sponsor carrying out the experimentation and the patient. Potentially significant impacts on the public are virtually ignored. No attempt is made to involve citizens at large beyond the general requirements of information and transparency. Even though the "role of the public" is repeatedly mentioned, the idea of "public awareness and understanding" and the necessity for a "public discourse" on XT are vague. Meetings held by the SACX between 2001 and 2005 (when the committee was discontinued), though officially open to the public, were oriented in practice toward experts and stakeholders. In a sense, the SACX itself can be seen as representing society at large through its hybrid composition (scientists, social scientists, philosophers, stakeholders, and an XT recipient) and of its capacity to interface between scientific and normative languages (for a parallel, see Jasanoff, chapter 5, this volume). Though recognizing that "public discourse on XT research is critical and necessary" and "public awareness and understanding of XT is vital because the potential infectious disease risks posed by XT extend beyond the individual patient to the public at large" (PHS 2001, 9–10), U.S. policy did not go beyond an abstract concept of the public (needing inputs and education), adequately and rationally represented by relevant expert committees.

Most of the precautionary measures designed to avoid threats to public health reside with the patient, who remains the key focus of the guideline. The technical provisions are tied to the principle of individual autonomy, whereby adult individuals in full possession of their faculties are presumed capable of giving their informed consent and are exclusively responsible as the only recipients of the possible benefits and risks of experimentation. Thus, where capacity exists, there is no need to involve third parties (family or close contacts) by asking them to be formally involved.

The U.S. guideline, then, essentially integrates the established approach to human subjects research with provisions on risks of contagious diseases. The weakness of this approach becomes apparent when we look at the context and the content of informed consent. The definition of XT itself provides a warning about the borderline identity of the XT patient, referred to as "XT product recipient." As the revised notion of XT includes "human fluids, cells, tissues and organs which have been in ex vivo contact with living nonhuman cells, tissues, organs" (PHS 2001, 1.2.21), the distance and the difference between the XT product and the XT product recipient becomes evanescent. A metaphysical mirage separates the ontologies of what is human and what is not, and of where the product ends and the recipient begins.

The extensive article dealing with the requirements for consent—describing a long list of constraints on behavior—provides a framework for taming the identity of the informed recipient. The patient's awareness of "the potential for infection with zoonotic agents" and "the potential for transmission to the recipient of unknown xenogeneic infectious agents" becomes a prelude to giving consent to the "potential need for isolation procedures during any hospitalization" and the "importance of complying with long-term or life-long surveillance" (PHS 2001, 2.5.5, 2.5.7). Liberty, in other words, will be willingly sacrificed by the properly educated patient.

In order to maximize the potentiality of informed consent, the guideline includes among patient's responsibilities the control of family members and close contacts. During the consent process, the patient is informed of the responsibility to educate close contacts regarding the possibility of xenogeneic infections (PHS 2001, 2.5.4) and of the "behavioral modifications" that entering the clinical trial will involve for them (e.g., duties of reporting any significant unexplained illness, limitations to unprotected sex, breastfeeding, and any activity that involves potential exchange of blood or body fluids). This extension of patients' obligations to the control of third parties, deriving from the imperfect analogy between XT and general principles for human experimentation, makes the limits of the U.S. model evident. Control of potential risks of infectious diseases can hardly be achieved through the structure of individual informed consent aimed at authorizing personal risks and constraints and not at controlling others.

After the publication of the draft guideline, a proposal was made to introduce informed consent for close contacts. The internal FDA debate made it clear that this new form of consent for close contacts was not a proxy consent, but rather notice of what they would be expected to do and might be required to do. Through this strategy, the basic formula of individual consent was never challenged.

What makes informed consent effective in this scenario is patients' acceptance of lifelong surveillance. Risks can be effectively isolated because, by accepting all containment measures, patients agree to be isolated from the rest of the world. In the context of XT, the act of consenting is just one aspect of a complex screening procedure: the patient's choice to undertake the trial is preceded by and subordinated to the investigators' selection of the patient. Only reliable and compliant patients are allowed to "choose" to undergo the experimentation.

The fact that patients are more chosen than choosing emerged in an early discussion of the safety measures to be adopted by FDA. In principle, people should not be socially discriminated against when suffering from addictions but, because of their presumed unreliability, they are de facto in an allotransplantation subgroup, and XT will give them even fewer chances to be selected as recipients: "Since those people are not generally considered suitable recipients of human organs, they are potentially a very desirable population. Nevertheless, this committee and others experts have told us these transplants should go into people that you can count on for the rest of their lives" (FDA 1997, 296).

But it is still science that provides the overall credibility of the individually based U.S. policy: XT science is not framed independently from but in accordance with the liberal model of behavior informing the guideline. The tacit assumption lying behind the construction of individual risk is that potential xenogeneic pathogens will be compatible with a social order in which scientifically informed individual choices may prevent major infections and epidemics. Although infections from XT might be unpredictable, they are assumed to be mostly controllable by individual good will and educated compliance.

The main analogy adopted by the guideline in dealing with risks of transmissible diseases is a scientifically well understood, socially organized, and familiar epidemic: AIDS. Though other diseases are occasionally mentioned in the guideline, HIV remains the paradigm for potential XT risks (FDA 1997, 291–292), as a "persistent latent infections may result in person-to-person transmission for many years before clinical disease develops in the index case, thereby allowing an emerging infectious agent to become established in the susceptible population before it is recognized" (PHS 2001, 15). But the numerous mentions of AIDS as the danger that XT clinical trials should avoid do not sound like a threat or warning. Instead, they function more as a reassuring scenario, reducing unknown risks to a mundane situation, a disease that can be adequately controlled through responsible behavior.

Certainly AIDS, as the major epidemic of the second half of the twentieth century, is a relevant model. At least some of its presumed causes are related to laboratory experiments (Elswood and Stricker 1994), and it took time for scientists to become aware of it and to alert, organize, and control the more exposed communities. All this knowledge constitutes an important learning experience about infectious diseases. At the same time, AIDS also reinforces the assumptions that epidemics can be managed through educating individuals and increasing their awareness, through adoption of a familiar frame of rights and responsibilities.

Yet the model of a manageable epidemic would be far less applicable to epidemics of SARS, Ebola, or avian flu, infectious diseases that are less susceptible to control and whose containment can be achieved only through people's containment and isolation. The unknowns about XT do not allow us to anticipate whether the former or the latter example is more pertinent.

Other less controllable infections requiring "the most restrictive level of isolation or isolation precautions" (PHS 2001, 38) are listed separately in the guideline, as requirements for infection control and not for informed consent. In fact they are not relevant to consent, as compulsory public health measures are not meant to be voluntary. However, the fear of epidemics has led to the introduction of an indirect strategy to expand the sphere of safety. In 2004, a rule concerning the eligibility of human cell and tissue donors excluded close contacts of xenotransplanted patients, labeling them as "close contact of a xenotransplantation product recipient." (PHS 2001, 1.3 and passim) The circle of connections through definitions grows. A chain of words implying a chain of relations—from the XT-product to the XT-product-recipient, to the XT-product-recipients' intimate contacts—suggestively conveys the idea that the boundaries of risk are blurred among cells, species, and people directly or indirectly involved in experimental treatment (DHHS 2004). Risk, in other words, subtly overflows the framework of individual informed consent.

The European Model: Waiving Rights for Innovation
In 1999, the Parliamentary Assembly of the COE unanimously adopted Recommendation 1399, which called for a legally binding moratorium on all XT clinical trials. Although some European countries adhered to the moratorium, the Committee of Ministers of the COE set up a multidisciplinary Working Party on XT (WPx) composed of specialists in ethics, law, medical research, clinical practice, epidemiology, immunology, and animal protection to prepare a report on the state of the art in the field of XT as groundwork for a new Recommendation.

A willingness to maintain the connections between the technical and the normative aspects of XT characterized this European approach (Tallacchini 2001). Solidarity was also present as a distinctive element of the

European vision of bioethics—as opposed to the individualistic American approach—as capable of organically connecting individuals and society at large. If the use of explicit ethical language in matters of science sounds improper from the perspective of U.S. policy (Jasanoff 2005), the European WPx was at ease in mixing descriptive and normative terminology even in addressing scientific facts—for instance, raising questions about the ethical status of microchimerism and the coexistence of human and porcine stem cells (COE 2000, 4.3.5).

This holistic approach deeply affected the links between individual consent, family consent, and acceptance by society. Each step was perceived as problematic because of restrictions on individual freedom, risks placed on health workers and family members, and the creation of new societal risks. Therefore, only a network of safeguards and social cohesion through the participation of all parties involved could make clinical experimentation acceptable. Informed consent by relatives was discussed not as a potential veto on patient consent, but instead as an element of a broad social negotiation on the acceptance of a high risk procedure.

The European document was supported by the conviction that technologies such as XT must have the approval of society, not merely the patient. Citizens have to agree in order for the risks of a new biotechnology to be legitimate, and institutions should actively contribute to creating and safeguarding initiatives of public importance. Despite this initial position, by the time the final report was completed and the new Recommendation 10 (2003) on XT was approved, deep changes had occurred in risk perception. Superficially, the main difference is that the 2003 report gives greater attention to public health. The final report noted that several national and international surveys had shown that XT posed more problems than expected, and it recalled how health emergencies taking place in Europe (namely BSE, Creutzfeldt-Jakob disease, and the anxiety over GM food) had "created an atmosphere of distrust of science and scientists in the public mind," so that "no one is really 'in control' or 'knows what will happen'" (COE 2003a, 53).

Although neither report was supportive of the precautionary principle, they both accepted the idea that "all those involved in deciding about XT must be satisfied that the risk to the individual recipient, their families, the medical and nursing teams, and the general public are minimal and controllable" (COE 2003a, 70). In practice, this meant that precaution against XT suddenly switched to precaution against xenotransplanted patients. The PP, initially applied to XT as a source of potential risk by Rec1399(1999), was redirected by Rec10(2003) against the patient as the most tangible and controllable source of risk.

Facing the indeterminate risks posed by XT, the COE did not reduce potential unknowns to familiar risks as the U.S. guideline did in order to maintain its focus on the liberal individual. Instead, the European approach considered the risks associated with XT so unpredictable that an authoritarian and compelling vision of the state seemed unavoidable and fundamental rights of human subjects had to be waived, as "it is hard to see how the normal pattern of a freedom to withdraw can be allowed" (COE 2003b, 69). The EU fully endorsed this approach to regulating XT. Commenting on the COE report, the European Commission Scientific Committee on Medical Products and Medical Devices accepted without hesitation that in XT some fundamental human rights—for the patient and others may be suspended. "As XT has implications for public health, it may be that certain rights may have to be modified in such a way that surveillance can be continued. Patients (and others?) could therefore have to agree to waive some of their human rights" (Scientific Committee on Medical Products and Medical Devices 2001, 11).

Though pointing out that its goal is "to protect all persons involved in XT . . . as well as public health" (Article1), Rec 10(2003) de facto reversed the prioritization of individuals over the state in biomedical matters, endorsing public health protection and innovation as main values, and subordinating informed consent to these aims. In this perspective, too, patients are chosen rather than choosing, as in the United States, but the overwhelming power exerted by medical personnel was also manifests itself in a blatant way. According to the report, "those engaged in transplant work stress that the recipients will be well known"—because their identities and their data will be thoroughly controlled—"and therefore it will be easy to assess how willing they will be to conform. Transplant patients are historically conformist" (COE 2000, 6.5.2). Noncompliant patients and personal contacts will be taken care of by public authorities through appropriate interventions, namely registration, compulsory medical follow-up, and sampling. Reversing the relationship between individuals and the public, the Recommendation replaced the concept of individual rights with that of protection—switching the patient from a subject of rights to an object of protection—thus legitimating compulsory constraints.

Though the European legal construction of XT seemingly violates the European Convention on Human Rights, when asked to give their advisory opinion, representatives of the European Court of Human Rights (ECHR) upheld the legitimacy of constraints. In explaining their reasons, ECHR's members even moved beyond the terminology of the Recommendation, making clear that the expression "compulsory constraints"—or interferences—means lawful "detention": "The Convention in Article 5(1)

(d) permitted the lawful detention of persons to limit the spreading of infectious diseases. . . . Interferences could be justified provided they were necessary in a democratic society. Moreover, in certain circumstances it might be considered that an individual, by giving consent to a particular interference, had waived his or her rights" (COE 2003b, Appendix, 34).

The ECHR opinion failed to recognize that XT is not a natural emergency of the kind that triggers public health compulsion, but a new technology searching for legitimacy. Accordingly, ECHR simply reinforced COE's position: "The surveillance procedures associated with XT can only be effective if they are complied with to the letter. . . . Many of the rights in the Convention [on Human Rights] were subject to permissible restrictions and involved establishing a proper balance between competing interests" (COE 2003b, 34).

Though Rec10(2003) emphatically ascribed an important role to the public, public opinion is subsumed by the activities of the regulatory bodies, advisory groups, and working parties who provide the framework within which proper debate may take place. What the Recommendation really meant by referring to the public dimension is the hold that the public health framework has acquired on the European governmental imagination—a hold that causes individual rights to fade away when confronted with the epidemiological perspective.

This concern for public health is something more than the state's responsibility for its citizens' lives in accordance with the classic terms of biopolitics. Though public health crises endangered the credibility of European science policy and created pressure for a dedicated plan to restore public trust, they also represented a powerful opportunity for Europe to project its image as a credible political entity capable of serving European citizens, even though this image was mainly projected on the scary landscape of risks (Jasanoff 2005).

In legal theory, H. L. A. Hart (1961) framed the survival of humankind as the "minimum content" of natural law, meaning that as a matter of fact—according to a positivist concept of the law—every legal system has to accept this minimal assumption if a lawful community is to exist at all. In fact, Hart says, this is "something presupposed by the terms of the discussion; for our concern is with social arrangements for continual existence, not with those of a suicide club" (Hart 1961, 188). In the EU context, safety, as a sanitized version of survival, has become the basic value around which the Union has construed its commercial-political identity—yet almost all other ethical values belong to and can be established by member states. The appeal to safety is a way to reinforce the construction of the political Europe, which widely refers to that mandate in order to legitimize its imagined collective identity (Tallacchini 2009; also Dratwa, chapter 12, this volume).

Canadian and Australian Models: Expert-Citizens and Watchdog Communities

In contrast to the American and European approaches, locked inside their traditional visions of government, Canada and Australia seem to have used XT as a case to test not only an experimental biomedical technology but also an experimental (and experiential) idea of democracy, explored through the involvement of professional and social communities, interest groups, and individual citizens, and through heterogeneous techniques aimed at generating new forms of decision making. Nonetheless, the theoretical frameworks lying behind the involvement of citizens in the deliberative process on XT were neither explicit nor unequivocal. Though assessing new ways to deal with complex policies, neither model was immune from seeking consensus and each to some extent reinvoked older communitarian beliefs.

A War over Publics

The argument for XT in Canada and Australia illustrates not only how theories and techniques of public involvement have changed over the years, but also how diverse interests and points of view have fought each other in order to have their image of the public emerge and win. The position toward public involvement adopted by a large fraction of the XT scientific community has, with very few exceptions (Ivinson and Bach 2002), been quite skeptical.

Beginning in the mid-1990s, when the idea of initiating clinical trials became realistic and the first Gallup poll reported public enthusiasm (Gallup/Partnership for Organ Donation 1993), a letter in *Nature* from a group of Australian researchers reported high rates of aversion to XT among 1,728 acute care nurses working in fifty-nine public hospitals in Australia. "A common assumption is that patients with end-stage organ failure will be enthusiastic participants in XT programmes. However, we report data that challenge this assumption" (Mohacsi et al. 1995, 434).

Two years later, while international discussion on XT was becoming more intense, the same researchers suggested that aversion to XT was also coming from a group of patients waiting for transplants, and that these negative attitudes needed to be considered before proceeding with XT (Mohacsi et al. 1997). Though the Australian data were never contested, most researchers in the field reacted quite resentfully, providing contradictory numbers and not reasons. Evidence that two of the most relevant communities touched by XT, namely patients and health care workers, might not be excited about XT was perceived as a threat by the scientific community and a challenge to the rhetoric of saving lives.

Soon after, a commentary in *Lancet* contrasted the critical Australian results with the enthusiastic support of UK patients. "I sent a questionnaire to over 850 dialysis patients known to the British Kidney Patient Association. . . . Although the 113 patients who were approached in Australia certainly did not give encouragement to the continuation of the programme, renal patients in this country are both enthusiastic and supportive" (Ward 1997, 1775). In response, Mohacsi ironically commented on "scientific enthusiasm" as an unusual criterion in the realm of scientific evidence (Mohacsi et al. 1998). The debate on public attitudes toward XT was instantly transformed into a war of numbers (850 against 113) and local attitudes (UK against Australia), rapidly intensifying in the following years, especially after the call for a moratorium.

Although worries about potential xenogeneic infections involving the public at large were not directly addressed, discussion in the XT scientific journals was also governed by the rhetoric of numbers: the large numbers of patients waiting for an organ, the small chances of providing them with a human organ, the numerous days of survival of xenotransplanted animals, and the numerical evidence that clinical trials were reliable. In the following years, as risks of infectious diseases began to appear as a major obstacle to XT, the attempt to display support from the public grew with a flurry of surveys. The number of analyses performed and the relevance that scientific journals started to attribute to the social assessment of XT have constantly risen. The space devoted by the specialized literature to quantitative studies of the socioethical issues of XT—with the simultaneous dismissal of qualitative studies as "often difficult to quantify" (Hagelin 2004) has dramatically increased. The impression that public acceptance of XT remains critical for its feasibility has also given rise to much advertising aimed both at strengthening patients' compliance and feelings of belonging to imagined XT communities (Novartis 2001), and at blurring the boundaries between swine and humans, depicting the pig as "one of us."

Until recently, most surveys produced by scientists involved in XT research have been built on the imaginary of the "deficit model," according to which the public does not understand science, people need education, and the more educated public is imagined as being more willing to accept new technologies (Wynne, Felt, et al. 2007). In XT, publics have been used as a multifaceted, strategic political tool, their role either affirmed or denied, recognized or dismissed, according to circumstances. It is in this context and in response to the growing idea of public involvement in science-based decisions, that the participatory initiatives of Canada and Australia are situated (Davies et al. 2003).

Canada

Starting in 1999, the Canadian government launched an extensive public program on XT culminating in the release in 2001 of the report *Animal-to-Human Transplantation: Should Canada Proceed? A Public Consultation on XT* (CPHA 2001). As the report explained, the government was not just seeking public approval on high-impact technologies, but also strengthening democracy in health policy (Tallacchini 2002). The initiative introduced an innovative role for citizens: that of the lay scientist. The Canadian initiatives imagined and shaped the public as composed of lay-experts and citizen-scientists: people capable of acquiring specific skills and knowledge about scientific issues in order to participate in decision making. Even though a relatively small fraction of the population was actively involved in the process, the experiment aimed to promote a lasting learning experience.

The final document produced through the consultation proposed a "made in Canada" approach to XT, strongly focused on how decisions had been shaped by information and how peoples' identities had been transformed by the learning process (CPHA 2001, 11). Special emphasis was placed on the metamorphosis of uninformed Canadians into informed Canadians, and to the unexpected outcome of this transformation. "When generally uninformed Canadians were asked if Canada should proceed with XT, the majority said yes. However, as they became more informed, a shift occurred and the majority of informed Canadians said no, Canada should not proceed" (CPHA 2001, 10).

The conclusion that "Canada should not proceed" was envisaged as a special form of knowledge, merging science with social wisdom. Reflecting on their new awareness, representatives of the Canadian public wished to know more both about science and about cooperative behavior, healthy lifestyles, and fair allocation of resources. The precautionary principle was invoked as part of a democratic call for a moratorium: informed Canadians were not opposed to XT in principle, but thought that those who wished to proceed with XT (the scientific community and industry) had to demonstrate the level of risk and how benefits would outweigh risks. Finally, these citizens asked for continued public discussion on XT.

The disappointing outcome of public consultation in Canada did not go unnoticed in the XT scientific community. Writing in two main journals in the field, a Canadian physician, James R. Wright sought to reinterpret the results of the Canadian report. Arguing that the actual attitudes of Canadians had been misrepresented and observing that the Canadian report should not be taken literally, Wright recalculated the numbers behind the consultation. Introducing a new ad hoc version of the public—namely, "meaningfully informed Canadians"—to contrast with the report's "informed Canadians," he argued that if the latter refused XT, the more reliable (and statistically recalculated) former group were in favor of it:

When presented (albeit unevenly) with the best available information on potential risks, benefits, and alternatives, as well as on the legal, social, and ethical implications of XT, 35% of citizen panelists voted "never proceed" to clinical trials, while 65% voted either "qualified yes" or "qualified no" (defined as "not yet"). As the difference between "qualified yes" and "not yet" is vague and probably highly dependent upon the group dynamics at each of the regional fora, I believe that these should be interpreted as essentially the same. This suggests that approximately two-thirds of "meaningfully" informed Canadians support XT. (Wright 2003, 476)

If the scientific community remained skeptical and disappointed, the Canadian government itself ended up adopting an unclear position. Though acknowledging the importance of Canadians' point of view, Health Canada established a Working Group to further analyze XT before adopting a policy (Health Canada 2001). Having launched the innovative, mostly qualitative initiative, the Canadian government remained puzzled about its political meaning, though it did de facto endorse the concept of meaningful public consultation. The easiest way for the Canadian government to deal with the challenges of its new democratic procedures was to step back toward the more familiar regulatory strategy of relying on expert advice.

Australia
In 2001, the Australian National Health and Medical Research Council (NHMRC) established a Working Party to provide advice on the scientific and normative aspects of XT, to produce guidelines on clinical trials, and to consult with the community. In the next two years, an informed community discussion was launched (NHMRC 2003a; NHMRC 2003b). In imagining its public, the Australian government introduced the two related identities of the socially responsible scientist and the responsible citizen

to deal with social risks, namely risks depending on the social ramifications of the technology. As the noncompliant patient is seen as the major source of risks, scientists and citizens were made responsible for patients' behavior in hospitals and in society.

In order to proceed with XT, both the U.S. and EU regulatory models asked for scientific criteria of safety and efficacy to be established, but did not engage in the social dimensions of XT except in terms of surveillance and containment. The Australian report, on the contrary, expressed some concern in the respect that scientific experimentation whose safety relies on patients' constraints is unfair, and science that accepts as one of its conditions the violation of fundamental human rights is not sound science. It is not prudent, the Australian documents explained, to assume that sick and vulnerable people will keep their word; nor is it ethical to require them to make such promises. Finally, it does not advance sound science to rely on a promise as an infection control measure (NHMRC 2003a). Compulsory measures should not be used as a surrogate for scientific evidence of safety; therefore, responsible investigators have to provide evidence of the safety of the proposed procedure and make restrictive monitoring or quarantine unnecessary. For all these reasons, "investigators should provide sufficient evidence of safety to show that there is no undue risk to the community if some participants choose to leave the trial" (NHMRC 2003a, 79).

Here the duet of the socially responsible scientist and citizen comes into play. The former has to ensure that no undue risk for the community will occur; the latter is shaped as the member of a responsible community dealing with the (few) citizen-patients who may not comply with safety measures. "While it is expected that, under these circumstances, most patients will adhere to the request for lifelong monitoring, individual recipients may withdraw their consent" (NHMRC 2003a, 82–83).

The noncompliant patient is psychologically excused and legally protected as a weak subject, but also socially alienated from the rest of the community. Although patients are granted the right to withdraw their consent, this situation is described as rare, not only because Australian citizens are reliable, but also because the social community will ostracize noncompliant patients. Within this communitarian vision, not much room remains for individual freedom. The relationships between scientists and citizens play out in a well-disciplined cooperative community in which social control is imagined to be stronger than legal sanctions.

The Australian guideline relies on patients' feelings of belonging, stating that having entered the trial, xenotransplant recipients also have a responsibility to comply with such provisions. Moral norms and social relationships ordinarily appear more flexible than legally binding rules, but in a self-policing communitarian context the reverse may be true. The guideline's title itself—*Guide for the Community*—called attention to the communitarian perspective pervading the Australian approach: a strong watchdog community can enforce the required behaviors even more effectively than a legal measure.

In a later document, the Australian regulation stepped back to stricter and more direct regulation of xenotransplanted patients. The patient still holds the right to withdraw from the trial but is not allowed to withdraw from the monitoring and follow-up associated with infectious diseases. In September 2004, the NHMRC decided to adopt a five-year moratorium on clinical research into animal-to-human whole organ transplants in Australia (NHMRC 2004). In this case as well, the participatory process and the community enabling perspective remained ambiguous and not fully endorsed by the government: legally enabling a self-monitoring community may have appeared an even riskier prospect than the risks of XT.

At the end of 2009, a short statement published on the NHMRC website explained the Australian government's postmoratorium attitude toward xenotransplantation (NHMRC 2009). Manifesting confidence in the recent scientific developments in the field, and providing evidence that "risks, if appropriately regulated, are minimal and acceptable," the NHMRC expressed confidence in global regulatory developments, namely, the Changsha Communiqué endorsed by the WHO (2009) and the guidelines developed by the European Medicines Agency—with no mention of the new U.S. documents on tissues and cells. Australia, it appears, is ready to proceed with xenotransplants, whatever the state of readiness of its patient population.

"The World Is My Patient": Normalizing Globalized XT

This chapter has illustrated four epistemic and social constructions developed to normalize XT as a biomedical practice. In different national democratic contexts, legitimacy was achieved through particular, culturally conditioned schemes of scientific and normative coproduction. XT offered the opportunity to rethink some fundamental rights of patients and risk-exposed citizens by experimenting with different visions of individuals in their relations both to experts, the social community and the state. In each case, however, preexisting normative assumptions clicked into play, constraining the very terms in which patient and community rights could be framed.

Let us end, however, with a look to the future. Globalized practices of clinical experimentation and major changes in XT techniques are rapidly modifying national normative settlements, as in the Australian case, and reframing the underlying rights and risks. "The world is my patient," observed an editorial in the official journal of XT (Ravelingien 2005), offering the metaphor of a medicalized global world, as yet unruled and searching for new forms of agreement in law and clinical practice.

Several factors could have stopped XT from developing further as a global therapeutic intervention, the most important being that several recent infectious diseases (SARS, avian flu, swine flu) have spread from other species (Jones et al. 2008). Nonetheless, XT has reemerged, both in the regulatory systems here analyzed and at the international level, as a fully normalized technology (WHO 2009; Cozzi et al. 2009).

The shift of scientific and regulatory focus from organs to cells in genomic medicine and the consequent reframing of XT as cell therapy and tissue engineering represent the single most important move toward building transnational legitimacy. With the blurring of "naturally occurring" and "technology-related" threats from zoonotic infections, xenotransplants are being perceived as less threatening, as they are but one among several potential globally dispersed sources for infectious diseases, yet are useful for therapeutic purposes beyond transplants. The scientific strategy of building a more manageable cellular ontology for XT, bypassing old concerns about whole organs and organisms, has been accompanied by the reshaping of the therapy's normative correlates. As this process has just begun, only a few sketchy reflections may be offered here.

Organs derived from human bodies—laden with ideas of subjectivity and personal identity—are tightly coupled to national social norms and regulations. Cells and tissues represent both a less ethically troubling (and thus more "neutral") framing for XT, and a more easily standardized and less invasive technology than organ transplants. Accordingly, whereas xenogeneic organs have been associated with national regulatory specificities, the reduced scale of xenogeneic cells—perceived as technological products—can be the object of an internationally shared regulatory approach. In this shift (from organs to cells, and from the national to the international), rights and risks will also necessarily be reshaped. As individual risks are imagined to be reduced and dispersed, individual rights are also being "resized" and minimized, as collective rights are reenunciated as detailed lists of safe behaviors. The precautionary principle as such has disappeared from policy agendas, and the "principle" of traceability—keeping track of all elements involved, from subjects to processes to products—has replaced precaution as the main legitimizing discourse on safety. In this process of regulatory normalization, XT has come to resemble ordinary forms of risk, associated with well-known and well-characterized hazards, rather than as a foray into the zone of biologically mysterious unknowns. A passage toward a new "politics of chimeras" and a new bioconstitutional moment is thus taking place in relation to XT, with the global technoscientific and regulatory orders mutually generating, or coproducing, each other.

References

ABC (Australian Broadcasting Corporation). 1995–1996. Xenotransplants. Making Animals of Ourselves. Retrieved from <http://www.abc.net.au/quantum/info/xt4.htm> (accessed November 2009).

Annas, J. George. 1985. Baby Fae: "Anything goes" school of human experimentation. *Hastings Center Report* 15 (1): 15–17.

Bach, Fritz H., Jay A. Fishman, Norman Daniels, Jenny Proimos, Byron Anderson, Charles B. Carpenter, et al. 1998. Uncertainty in XT: Individual benefit versus collective risk. *Nature Medicine* 4:141–144.

Bach, Fritz H., and Harvey V. Fineberg. 1998. Call for moratorium on xenotransplants. *Nature* 391:326.

Bailey, Leonard L., Sandra L. Nehlsen-Cannarella, Waldo Concepcion, and Weldon B. Jolley. 1985. Baboon-to-human cardiac XT in a neonate. *Journal of the American Medical Association* 254:3321–3329.

Bailey, Leonard L. 2005. Candid observations on the current status of xenotransplantation: Invited commentary. *Xenotransplantation* 12:428–433.

Chapman, Louisa E., Thomas M. Folks, Daniel R. Salomon, Amy P. Patterson, Thomas E. Eggerman, and Philip D. Noguchi, et al. 1995. Xenotransplantation and xenogeneic infections. *New England Journal of Medicine* 333:1498–1501.

COE (Council of Europe) Working Party on Xenotransplantation. 2000. State-of-the-art report on xenotransplantation. Strasbourg, July 7; Interim report on the state of the art in the field of XT. Strasbourg, October 25.

COE (Council of Europe) Working Party on Xenotransplantation. 2003a. Report on the state of the art in the field of xenotransplantation. Strasbourg, February 21.

COE (Council of Europe) Working Party on Xenotransplantation. 2003b. Explanatory memorandum to Recommendation 10 (2003) of the Committee of Ministers to member states on xenotransplantation. Strasbourg, June 5. Retrieved from <http://www.coe.int/t/dg3/healthbioethic/Activities/06_Xenotransplantation_en/INF_2003_12exenoER.pdf> (accessed December 2010).

CPHA (Canadian Public Health Association). 2001. Animal-to-human transplantation: Should Canada proceed? A public consultation on xenotransplantation. Retrieved from <http://www.xeno.cpha.ca/english/index_e.htm> (accessed November 2009).

Cooper, David K. C., and Robert P. Lanza. 2000. *Xeno: The promise of transplanting animal organs into humans.* Oxford: Oxford University Press.

Cozzi, Emanuele, Mariachiara Tallacchini, Enda B. Flanagan, Richard N. Pierson III, Megan Sykes, and Harold Y. Vanderpool. 2009. The International Xenotransplantation Association consensus statement on conditions for undertaking clinical trials of porcine islet products in type 1 diabetes. *Xenotransplantation* 16:203–214.

Davies, Gail, Jacquie Burgess, Malcolm Eames, Sue Mayer, Kristina Staley, Andy Stirling et al. 2003. Deliberative mapping: Appraising options for addressing "the kidney gap." Wellcome Trust. Retrieved from <http://www.deliberative-mapping.org> (accessed November 2009).

DHHS (Department of Health and Human Services). 2004. Eligibility determination for donors of human cells, tissues, and cellular and tissue-based products. *Federal Register* 69 (101): 29786–29834.

DHHS-SACX (Department of Health and Human Services, Secretary's Advisory Committee on Xenotransplantation). 2004. Informed consent in clinical research involving xenotransplantation (draft). Retrieved from <http://www.transplantation-soc.org/downloads/SACX-informed-consent.pdf> (accessed August 2009).

Dorling, Anthony. 2002. Clinical xenotransplantation: Pigs might fly? *American Journal of Transplantation* 2:695–700.

EMA (European Medicines Agency) Committee for Human Medicinal Products (CHMP). 2009. Guideline on xenogeneic cell-based medicinal products. EMEA/CHMP/CPWP/83508/2009. London, February 27. Retrieved from <http://www.emea.europa.eu/pdfs/human/cpwp/8350809en.pdf> (accessed November 2009).

Elswood, B. F., and R. B. Stricker. 1994. Polio vaccines and the origin of AIDS. *Medical Hypotheses* 42:347–354.

FDA (U.S. Food and Drug Administration). 1997. XT Subcommittee/biological response modifiers advisory committee/center for biologics evaluation and research. Meeting transcript, December 17. Retrieved from <http://www.fda.gov/ohrms/dockets/ac/97/transcpt/3365t1.pdf> (accessed November 2009).

FDA (Food and Drug Administration), DHHS (Department of Health and Human Services), PHS (Public Health Services). 1996. *Draft public health service guidelines for xenotransplantation. Federal Register* 61 (185): 49919–49932.

Fishman, Jay A. 1998. Letters to the editor. *Nature Medicine* 4 (1): 372.

Gallup/Partnership for Organ Donation. 1993. *The American public's attitudes toward organ donation and transplantation.* Boston: The Partnership for Organ Donation

Groth, Carl G., and Olle Korsgren. 2008. Proceedings of "pig-to-man islet transplant summit" held at the Nobel Forum, Stockholm, June 4–5, 2007. *Xenotransplantation* 15:77–78.

Hagelin, Joakim. 2004. Public opinion surveys about xenotransplantation. *Xenotransplantation* 11:551–558.

Hart, Herbert L.A. 1961. *The concept of law*. Oxford: Oxford University Press.

Health Canada. 2001. Revised fact sheet on xenotransplantation. Retrieved from <http://www.hc-sc.gc.ca/dhp-mps/alt_formats/hpfb-dgpsa/pdf/brgtherap/xeno _fact-fait_e.pdf> (accessed November 2009).

Ivinson, Adrian J., and Fritz Bach. 2002. The xenotransplantation question: Public consultation is an important part of the answer. *Canadian Medical Association Journal* 167:42–43.

Jasanoff, Sheila, ed. 2004. *States of knowledge: The co-production of science and social order*. London, New York: Routledge.

Jasanoff, Sheila. 2005. *Designs on nature: Science and democracy in Europe and the United States*. Princeton: Princeton University Press.

Jones, Kate E., Nikkita G. Patel, Marc A. Levy, Adam Storeygard, Deborah Balk, John L. Gittleman, et al. 2008. Global trends in emerging infectious diseases. *Nature* 451:990–993.

Kushner, Thomasine, and Raymond Belliotti. 1985. Baby Fae: A beastly business. *Journal of Medical Ethics* 11:178.

McLean, A. M. Sheila, and Laura Williamson. 2005. *Xenotransplantation: Law and ethics*. Aldershot, UK: Ashgate Publishing.

Mohacsi, Paula J., Charles E. Blumer, Susan Quine, and John F. Thompson. 1995. Aversion to xenotransplantation. *Nature* 378:434.

Mohacsi, Paula J., John F. Thompson, J.K. Nicholson, and D.J. Tiller. 1997. Patients' attitudes to xenotransplantation. *Lancet* 349:1031.

Mohacsi, Paula J., John F. Thompson, and Susan Quine. 1998. Attitudes to xenotransplantation: Scientific enthusiasm, assumptions and evidence. *Annals of Transplantation* 3:38–45.

Najarian, John S. 2003. Experimental xenotransplantation: A personal history. *Xenotransplantation* 10:10–15.

NHMRC (National Health and Medical Research Council), Xenotransplantation Working Party. 2003a. Animal-to-human transplantation research: How should Australia proceed? Response to the 2002 public consultation on draft guidelines and discussion paper on XT, Commonwealth of Australia. Retrieved from <http:// www.nhmrc.gov.au/_files_nhmrc/file/publications/synopses/e54.pdf> (accessed August 2009).

NHMRC (National Health and Medical Research Council), Xenotransplantation Working Party. 2003b. Animal-to-human transplantation research: A guide for the community. Public consultation on XT 2003/04, Commonwealth of Australia. Retrieved from <http://www.nhmrc.gov.au/_files_nhmrc/file/publications/synopses /e54.pdf> (accessed November 2009).

NHMRC (National Health and Medical Research Council). 2004. Communiqué from the NHMRC's 154th Session. Perth, September 20. Retrieved from <http:// web.archive.org/web/20050620070744/http://www.nhmrc.gov.au/media/rel2004 /xenocom.htm> (accessed November 2009).

NHMRC (National Health and Medical Research Council). 2009. NHMRC statement on xenotransplantation. December 11. Retrieved from <http://www.nhmrc

.gov.au/media/noticeboard/notice09/091210-xenotransplantation.htm> (accessed July 2010).

Novartis. 2001. Transplant Square. Retrieved from <http://web.archive.org/web/20010201070400/http://www.transplantsquare.com> (accessed November 2009).

Nuffield Council on Bioethics. 1996. *Animal-to-human transplants: The ethics of xenotransplantation.* London.

Paradis, Khazal, Gillian Langford, Zhifeng Long, W. Heneine, P. Sandstrom, W. M. Switzer, et al. 1999. Search for cross species transmission of porcine endogenous retrovirus in patients treated with living pig tissue. *Science* 285:1236–1241.

Patience, Clive, Yasuhiro Takeuchi, and Robin A. Weiss. 1997. Infection of human cells by an endogenous retrovirus of pigs. *Nature Medicine* 3:282–286.

PHS (Public Health Service). 2001. Guideline on infectious disease issues in xenotransplantation. January 19. Retrieved from <http://www.fda.gov/Biologics BloodVaccines/GuidanceComplianceRegulatoryInformation/Guidances/Xeno transplantation/ucm074727.htm> (accessed December 2010).

Ravelingien, An. 2005. The world is my patient. *Xenotransplantation* 12:88–90.

Reemtsma, Keith, B. H. McCracken, J. U. Schlegel, M. A. Pearl, C. W. Pearce, C. W. DeWitt, et al. 1964. Renal heterotransplantation in man. *Annals of Surgery* 160:384.

Scientific Committee on Medicinal Products and Medical Devices, European Commission Health & Consumer Protection Directorate General. 2001. Opinion on the state-of-the-art concerning xenotransplantation. Retrieved from <http://ec.europa.eu/health/ph_risk/committees/scmp/documents/out38_en.pdf> (accessed November 2009).

Selgelid, Michael J. 2009. Pandethics. *Public Health* 123 (Special Issue): 255–259.

Starzl, Thomas E., T. L. Marchioro, G. N. Peters, C. H. Kirkpatrick, W. E. C. Wilson, K. A. Porter, et al. 1964. Renal heterotransplantation from baboon to man: experience with 6 cases. *Transplantation* 2:752–776.

Tallacchini, Mariachiara. 2001. Commentary: Council of Europe Working Party on xenotransplantation: State-of-the-art report on xenotransplantation (2000). *Xenotransplantation* 8:154–156.

Tallacchini, Mariachiara. 2002. Commentary: Regulatory issues in Europe and Canada. *Xenotransplantation* 9:371–373.

Tallacchini, Mariachiara. 2009. Governing by values. EU ethics: Soft tool, hard effects. *Minerva* 47 (3): 281–306.

Vanderpool, Harold Y. 2007. Informed consent in clinical research. *Xenotransplantation* 14:353–354.

Wallis, Claudia. 1984a. "Baby Fae stuns the world." *Time* 124:70. Retrieved from <http://www.psy.vanderbilt.edu/courses/hon182/Baby_Fae.pdf > (accessed December 2010).

Wallis, Claudia. 1984b. "Baby Fae loses her battle." *Time* 124:88. Retrieved from <http://www.time.com/time/magazine/article/0,9171,927010,00.html> (accessed November 2009).

Ward, Elizabeth. 1997. Attitudes to xenotransplantation. *Lancet* 349:1775.

Weiss, Robin A. 2000. Certain promise and uncertain peril. The debate on Xeno-transplantation. *EMBO Reports* 1:2–4.

WHO (World Health Organization). 2009. Changsha Communiqué: First WHO Global Consultation on Regulatory Requirements for Xenotransplantation Clinical Trials. Changsha, China, November 19–21. Retrieved from <http://www.who .int/entity/transplantation/xeno/ChangshaCommunique.pdf> (accessed November 2009).

Wright, R. James, Jr. 2003. Letters to the editor. *Xenotransplantation* 10:475–476.

Wynne, Brian, Ulrike Felt, Michel Callon, Maria Eduarda Gonçalves, Sheila Jasanoff, Maria Jepsen, et al. 2007. Taking European knowledge society seriously. Commission of the European Communities, Brussels. Retrieved from <http://ec.europa.eu/research/science-society/document_library/pdf_06/european -knowledge-society_en.pdf> (accessed August 2009).

Two Tales of Genomics: Capital, Epistemology, and Global Constitutions of the Biomedical Subject

Kaushik Sunder Rajan

In this essay, I present two case studies of how genome science emerged and touched down in the early 2000s. One case is from the United States and the other from India (Sunder Rajan 2006). I develop these cases in the context of this volume in the hope that their juxtaposition to the other contributions will bring to light certain questions about rights, subjectivity, and subject-constitution in the life sciences. I am particularly interested in two questions:

How might we complicate an understanding of subjectivity as it emerges and comes to be at stake in contemporary global biomedicine?

What does this emergent subjectivity have to do with the global political economy of the life sciences as related to therapeutic development?

First, I wish to say something about why subjectivity is a critical site at which one can explore questions of relevance to this volume. Our concerns with "reframing rights" are not abstract normative or moral questions of what rights *should* be recognized in an era of emergent biotechnologies. They are, rather, concrete and empirical: *how* are rights actually being reframed in the emergent coproductions of law and life sciences? Bioconstitutionalism arose as an animating conceptual frame for this volume in the course of conversations among the contributors, to accommodate their empirical findings. Consistent with my ongoing research interests, I have attempted to explicate the place of capital and its relationship to global neoliberal formations in these conversations.

Bioconstitutionalism encompasses, I would argue, to the articulated and constantly emergent relationship between knowledge and value. This relationship—which I would write as Knowledge/Value—is akin to the relationship of Power/Knowledge posited by Michael Foucault (Foucault 1980). Sheila Jasanoff (chapter 1, this volume) rightly points to Foucault's influential concepts of biopower and biopolitics in helping us conceptualize the ways in which life became a particular subject of political understanding in modernity. Crucially, both biopower and biopolitics require an understanding of the epistemological; further, an understanding that refuses to assume that knowledge about life can emerge independent of sociopolitical histories and contexts, such as those of nation states.

Knowledge/Value, like Power/Knowledge, raises questions concerning each term in the dyad—what is knowledge, what is value, and how do the two interrelate in the context of contemporary emergences in the life sciences, as well as in forms of governance? But it is also a question of the slash ("/") in between. This slash, I would argue, signifies coproduction: one cannot understand the production of knowledge or of value independent of the other.

Though I am indebted to Foucault in thinking about Knowledge/Value, I want specifically to emphasize the complicated dialectics that animate the mutual articulation of these two terms. Law and the life sciences are identified in this volume as key productive sites at which emergent forms of knowledge and value connect. What is at stake, then, empirically and conceptually, is to study the *modes and relations of production (and coproduction)* of knowledge and value at exemplary sites. These sites may be disciplinary (new subfields of the life sciences, for instance); institutional (places where research is performed, or laws are made, or, as in this chapter, value is created); or performative (where coproduced forms of knowledge and value are staged and displayed in choreographed ways). It is because of the importance of modes and relations of production (and coproduction) in biotechnology that I turn to Karl Marx as an essential supplement to Foucault. This turn to Marx, and to value as something that is coproduced with knowledge just as power is, allows a further conceptualization of biopower and biopolitics in terms of what I and others have referred to as *biocapital* (Sunder Rajan 2006).[1]

In reading Marx, it is worth remembering that his analysis of modes of production is not simply mechanistic. Indeed, Marx's understanding of the functioning of capital is always in terms of modes *and relations* of production; production, for him, is always a relational category—one that requires agency, but which also subjects people to its logics. Hence, subjectivity is implicitly a central conceptual category for Marx. Given the particular modes of industrial capitalist production that preoccupied Marx, the form of subjectivity that he was most immediately concerned with was labor. But what forms of subjectivity are consequent on which political economic developments is a historical and empirical question. Although it is possible to read Marx reductively, as concerned only with labor and the politics of the working class, it is important to understand that philosophically, Marx was trying to work through the broader question of the relationship between emergent modes of (co)production under particular technical, political, economic (and indeed epistemological) formations, entailing emergent forms of individual and collective subjectivity and sociality. That philosophical stance, and the analytic method it calls for, is of considerable relevance in understanding contemporary formations of biocapital.

The question of subjectivity as it emerges through genomics and its political economic context in different locales is therefore important because it speaks directly to one of the framing concerns of this volume: bioconstitutionalism (Jasanoff, chapter 1, this volume). It is important to think of subjectivity in its dual registers. Philosophically, the notion of subjectivity has been vexed. In its Hegelian formulation, it speaks both to *subjection* and *subjectivation*, that is, both to how one is subjected to something, and how one becomes the subject (agent) of it (Hegel [1807] 1977).[2] Anthropological discussions of subjectivity often focus upon the agential nature of the subject; indeed, work in anthropology and science and technology studies (STS) that has focused on the biosocial, or biological citizens, has tended to emphasize the self-molding and creative self-reinvention that occur in response to new biomedical knowledge, even when the subjects of this knowledge are disciplined, exploited, or suffering in various ways (for biosociality, see Rabinow 1992, 2007a; for biological citizenship, see Petryna 2002; Novas and Rose 2004; Rose 2006). In contrast, I am interested here in *subject-constitution*, the "always already" created subject whose agency is structured in culturally and historically specific ways. This subject may be constituted by many things (the dominant economic order being one of these) but, crucially, the subject is also constructed in a relationship to the state as well as capital—and thus is constructed in potentially very different ways in the emergent relationships between state and capital in different regions of the world.

This observation speaks to the multiple inflections of the term "constitution" as it appears in this volume. First, as a noun, constitution refers to a set of institutionalized codes, both legal and normative, that get sacralized (usually within a nation state) and held up as defining prescribed codes of action and governance. As noted in chapter 1 of this volume, the genomic text is in this expansive sense a constitutional text, no less than a conventional text in law. The working draft sequence of the human genome, announced in June 2000, was a revolutionary moment, not just for biology but for society and for the way in which "life" was configured in the sociopolitical imaginary. The sequence itself was referred to during its generation as the "Holy Grail" of life itself, and within the life sciences, it was seen at the time as constitutive of the ways in which research agendas on the nature of life would thenceforth be structured, conceptually, methodologically, and institutionally.

Second, "constitution" as a verb refers to the act of constituting, of putting in place. Work in science and technology studies has constantly insisted upon the constructed and contingent nature of production of scientific facts and technological systems. Recent literature on coproduction has further emphasized how scientific fact neither is produced in isolation, nor precedes or follows institutional productions (legal, corporate, governmental) in any causal manner (Jasanoff 2004). Rather, emergent science and technology are constituted along with emergent institutional structures and emergent individual and collective identities. This constitution requires active work by various actors whose goals are by no means identical or even coterminous. Tracing the acts of constitution that create postgenomic life across multiple locales and in multiple institutional settings is an integral part of any larger project that seeks to make sense of, and historically situate, the emergent life-worlds of the present.

Third, "constitution" refers to health, to the body, and its overall well-being. Indeed, genomics is about health, the body, life and hope, as it shapes and gets shaped by a society that is increasingly dedicated to therapeutic ends. This therapeutic society, like any other, has a political economy embedded within it. The modern bioconstitutional self/subject does not have complete freedom to pattern however she or he wishes in relation to biological discoveries, inventions, characterizations, and classifications. Just as there are multiple inflections on the term "constitution," so too there are multiple and complementary registers at which the bioconstitutional can be seen to emerge and play out. In this essay, I attempt to show how the neoliberal state configures the possibilities for that subject in ways that are very different in India and the United States. In this way, this chapter resists universalizing accounts of neoliberalism, globalization, and value itself.

I explore the American case through an account of Genomic Health, a biotechnology company in the business of "consumer genomics." I suggest that in this case, genomics—promising personalized medicine—configures subjects as *sovereign consumers*. I explore the Indian case through an account of Wellspring Hospital, with which were associated a genome startup called Genomed and a clinical research organization (CRO) called Wellquest. In this case, we see the configuration of genomics not with consumers but with *experimental subjects*. Adopting subjectivity as a critical site for analyzing locally specific histories of biocapital and bioconstitutionalism, I argue that Genomic Health and Wellspring are exemplary sites where one can empirically locate such an analysis.

A short note, then, about why, and in what ways, the two case studies are exemplary of genomics in the two locales. They are not exemplary in that they are necessarily "central" to genomics in either the United States or India, or even because they are representative of genomics research or corporatization in either country. Rather, they are exemplary because in each of these sites one can see the implosions of particular, consequential histories that must be understood if one is to understand global biomedical science and global capital today. These are two contingent sites that I, as an ethnographer embarked on a project of multisited ethnography, happened to study in the course of my research. But drawing them together and comparing them allows, hopefully, the restoration of certain kinds of historicity to processes of biocapital formation. My case study of Genomic Health, for instance, makes arguments that are not dissimilar to those of scholars such as Nikolas Rose about the biopolitics of life in neoliberalism (Rose 2006; though I do distinguish my analysis of the biopolitical from Rose's). But when locating this case next to the case of Wellspring, one is forced to see the ways in which a seemingly unmarked analysis of neoliberalism in fact is located within deep colonial histories and postcolonial inequities.

This essay, therefore, is a comparative analysis of subject-constitutions through genomics in the United States and India, though I do not wish to make each example stand in for genomics in its entirety in each nation. Rather, I am interested in following epistemology and technology together in the context of regimes of value generation in biomedicine. In the process, certain types of materializations occur as my cases illustrate. Why these materializations, why not others, and what others might possibly occur are serious questions for what Paul Rabinow refers to as an "anthropology of the contemporary" (Rabinow 2007b). Even though these individual cases cannot satisfy a full-blown project of symmetrical comparison, each points to and represents a certain differential in the social, political, and institutional national terrains. I hope that the comparative juxtaposition of the two cases allows a certain necessary historicization of global neoliberal forms and processes.

Genomic Health

Let us move now to the first case. Before I narrate my account of Genomic Health, it is important to situate the moment that I am writing about. My earlier accounts of Genomic Health (Sunder Rajan 2005, 2006) were written (if not published) at around the time that the company was being formed. At that time, I did not have the benefit of a certain historical distance that is already allowed now, a few years later. This is not to say that I am in a position to evaluate Genomic Health as a company, or even to situate its trajectory. It is, rather, to suggest that a company like Genomic Health was a marker of a particular moment in the history of genomics, and locating that moment is important.

Genomic Health was formed shortly after the working draft sequence of the human genome was generated. This moment in genomics was marked by what we might plausibly see as a transition in the genomic imaginary from informatics to therapeutics. In the late 1990s, as the race to sequence the human genome reached its peak, the key problem was the generation, storage, annotation, and classification of information. This was reflected in the business models of the most prominent genome companies of the time, which were primarily database companies. Their business models were based on their informatics capabilities, and their source of value often came from their databases, or from other technologies that could produce biologically meaningful data. This model became much less valuable after the completion of the working draft sequence of the human genome for three reasons. First, by this time a lot of information was already in the public domain, both because of public sequencing efforts and because of strategic attempts by academic and corporate actors to prevent the "locking up" of data through private ownership (see especially Sunder Rajan 2006, chap. 1). Second, the most useful commercial databases had already been bought by potential clients, so the ability to keep making money from these databases was limited. Third, the scientific and corporate imaginary of genomics had already shifted to "postgenomics": not just a question of what information resides within genomes, but what one might *do* with that information in scientifically and commercially valuable ways.

Further, the promise of genomics was that diagnostic or therapeutic tools could be tailored to individual genetic profiles, giving rise to personalized medicine. What personalized medicine actually meant was, in the early 2000s, largely in the realm of speculation. But it was central to the mantra of the genomic business of the time. It is in this context that one must locate the dawn of "consumer genomics," which is what a company like Genomic Health purveys.

Personalized medicine gets to the heart of "postgenomic" drug development. It refers to much more than specific therapeutic molecules (indeed, the generation of such therapeutics is rather difficult to achieve in practice for various biological reasons), imagining instead an entirely new ensemble of techniques, practices, and institutional structures of medicine. Steps involved in what might be called a "genetic approach" to the diagnosis and treatment of disease consist of the following (Collins and McCusick 2001): first, the identification of a disease with a genetic component; second, the mapping of the gene(s) involved in the disease to specific chromosomal regions; and third, the identification of the involved gene(s). At this point, one could develop diagnostics to identify the presence or expression of the involved gene(s) in patients to determine predisposition to the disease in question; use the gene itself as a drug (gene therapy); or understand the underlying biology of the disease to "rationally" develop therapeutics to target the molecular mechanisms of disease etiology. The diagnostic tests could further be a precursor either to steps taken to prevent the onset of disease (either interventionist or involving lifestyle changes), or to what is known as *pharmacogenomics*, which involves tailoring prescriptions and drug regimens to individuals based on their likely, genetically determined, response to these drugs. Some of these are easier to realize than others. For instance, it is much easier to use genomic information to create molecular diagnostic tests (including pharmacogenomic tests) than therapeutic molecules, because diseases are always complicated multifactorial events that are difficult to understand at the molecular level, and not necessarily easy to target and set right even if ever properly understood.

The epistemic and technological possibility of personalized medicine, which became more evident once the working draft sequence of the human genome was completed in 2000, led to the creation of new business models. One of the earliest companies to build its business model on personalized medicine was Genomic Health, cofounded by Randy Scott. Scott is a leading U.S. genome entrepreneur, who earlier founded Incyte Genomics, one of the major bioinformatic database companies, in the second half of the 1990s.

In his pitches for Genomic Health soon after its founding, Scott divided the "history" of genomics into decades: the 1980s, when the first companies brought products into the market based on classical DNA efforts; the 1990s, an era of industrialization and high-throughput technologies;

and the 2000s, which he called the era of consumer genomics, in which biological information merged with Internet capabilities have become key. In other words, the driving assemblages of emergent, "postgenomic" medicine, in Scott's opinion, are biological information (such as diagnostics), communications technologies that mediate such information as they travel to lay patients (the Internet), and the consequent networking of biosocial communities such as patient advocacy groups.

Genomic Health's vision, according to Scott, is

to build a new arm to the healthcare system. New genomic technologies will enable the world to characterize every patient's disease and health status as a complete genomic package. Every disease has a molecular basis and some level of genetic-encoded response. Individuals respond to therapy based on the molecular alterations of their disease and their own genetic code. Genomic Health's mission is to one day provide physicians and patients with an individualized molecular analysis that enables the treatment team to utilize relevant treatment guides for all diseases. Our ultimate goal is to make personalized medicine a reality and to dramatically improve patient care.[3]

This vision of consumer genomics is completely entwined with Scott's vision of Genomic Health as a company. He sees patients directly buying access to information and technologies, thereby seriously decentering the role of physicians. Pharmaceutical companies, in this imaginary, are still involved in therapeutic development, but Scott realizes that the lag between the development of diagnostic and therapeutic capabilities creates a window that is absolutely critical to the existence of Genomic Health as a company.

Scott accordingly outlines his notion of consumer genomics as follows:

Genomics is inherently personal. This is not about big industrial units that are bringing out products for other companies. Every one of us sits here with a genome, and a genome is our own particular story. My family has its story . . . they were a perfectly normal family; they thought they had no genetic disease, no genetic defects. . . . [But] no matter how healthy we may think we are, ultimately we will all face the reality of our genetic faults, and the diseases that are coming at us in the future. So we are all in this together.[4]

This quote encapsulates Scott's own visionary autobiography, superimposed upon the history, as he sees it, of genomics. In the process, Scott conjures up the image of a community, the community of patients-in-waiting, who are always already consumers-in-waiting. For lest we forget this voyage is overdetermined by the market, Scott immediately proceeds: "So the issue is how we bring these products to the market, how we bring them to the consumer."[5]

Scott's conceptions of consumer genomics reflect what Patrick O'Malley (2000, 465) describes as an emergent "enterprising prudentialism." O'Malley says, "The subjects of this technology of risk are imagined as *consumers* (albeit 'sovereign consumers'), for, as elsewhere in discourses of the freedom of choice, their liberty exists in the capacity to choose rationally among available options and to assemble from these the risk-minimizing elements of a responsible lifestyle" (465). Though this "liberty" is inherent to the very rationale of personalized medicine as a practice that is, in the first instance, *preventive*. This key addition that Scott makes explicitly forms the basis of potential value for Genomic Health. In this medical model, risk minimization and prevention are not dictated by the discriminatory practices of employers or health management organizations, or by the expert (and thereby, it is implied, forcible) interventions of physicians, but by patients themselves, who necessarily have to be configured as rational actors, capable of informed choice, in the way that direct-to-consumer advertising conceives of them. Diagnostic tests and preventive or therapeutic options, in the business models of personalized medicine, become consumables in exactly the same way as soap or perfumes. Genomic technologies, in the U.S. context, thus create "free" subjects of uncertainty (in the utilitarian sense of having rational choices of self-governance), who get subjected to a rationality of perpetual possible consumption, a rationality that, in a continually reenacted cycle, simultaneously demands "rational" self-governance and a governance that itself gets effected through further consumption.

The story of Genomic Health, I believe, provides insights into American neoliberal high-tech logics, which are based in what might be called a culture of innovation, driven in large measure by value systems of speculative capital, and a particular celebration of consumer culture. In Scott's articulation, the imagined sovereign consumers can exercise bottom-up choice in order to configure the market. This is precisely the imagination of sovereign subjectivity and autonomous rationality that underlies—and, in my view, complicates—ideas of biosociality or biological citizenship. It is the imagination of the market that leads to the configuration of subjects as sovereign consumers in the first place. Hence, although at face value the story of Genomic Health is suggestive of the sorts of subjectivity that Nikolas Rose (2006) diagnoses as constitutive to neoliberalism, the crucial point that I wish to emphasize is how immediate and seamless is the conflation of the imagination of a patient with that of consumer in the minds of actors such as Randy Scott. The consumer who exercises "free" choice in the genomic marketplace does not really have the choice whether to be configured as a consumer in the first place. Further, though the logic of personalized medicine that focuses on prevention might at first seem biopolitical in a Foucauldian sense (by focusing on care of the population and therewith care of the self), the imagination of the patient as always already consumer, and therefore always already a source of potential market value, complicates this reading. The insertion of consumer genomics into contemporary pharmaceutical value generation, as Joseph Dumit argues, is not so much about care of the population as it is about the generation of "surplus health."[6]

Genomic Health exemplifies not only the technical, epistemic, and market emergences that occur through genomics, but also the potential of genomics to constitute subjectivity in a particular market context. It is exemplary not because it is the dominant business model for consumer genomics or the one most likely to succeed in the market. Instead, it is exemplary because it brings together in one site a number of the key assemblages that have constituted genomics over the past decade. These include the informatics capabilities that lie at the heart of genomic technologies; the diagnostic potential that can be through genomics; the networking and communication possibilities enabled by the development of the Internet; the entanglement of the market in the development of the life sciences; and the larger structural logic of neoliberalism as constituted in the United States over the past thirty years, a time frame that coincides with the development of the biotech industry.

What interests me, however, are not just these dynamics in and of themselves, but the ways in which they are particular to the U.S. context, and are not replicated for instance in India, in spite of the India's self-conscious attempts to imitate an American "culture of innovation" in setting up its own biotechnology and genomics infrastructure. I turn next to the Indian experimental subject to juxtapose against my account of Genomic Health and the configuration of consumer subjectivity through genomics in the United States.

Wellspring Hospital

Wellspring Hospital was established by the Indian pharmaceutical company Nicholas Piramal India Limited (NPIL) in Parel, in the heart of the mill district of Bombay. In its initial conception, it housed Genomed, a genome startup seeded jointly by NPIL and the Centre for Biochemical Technology (CBT), India's flagship public sector genome lab, and also

Wellquest, a for-profit CRO set up to provide outsourced clinical trial services to biotechnology and pharmaceutical companies. The institutional structure of Wellspring-Genomed-Wellquest, in which a hospital, a genome company, and CRO are physically located in the same premises, itself suggests the threefold way in which genomics is imagined in the Indian context.

First, Genomed was set up in the context of the Indian state paying serious attention to genomics. This involved retooling one of India's national laboratories, the CBT, into the nation's flagship public sector genome lab.[7] Genomed was a startup seeded by the Indian state, in association with the investment fund The Chatterjee Group (TCG). One wing of Genomed was housed in CBT, in Delhi, and focused on research into the genetics of diabetes, schizophrenia, and asthma. The second was housed at Wellspring, in Bombay, and focused on pharmacogenomics. Although pharmacogenomics can be said to be an example of personalized medicine, it is not centrally concerned with developing genetics-based therapeutics, but rather with the testing of drug responses based on individual genetic profiles. In other words, pharmacogenomics lends itself to being incorporated into clinical trials to test the safety and efficacy of new drug molecules that are under development. Already, therefore, Genomed is located in Wellspring as part of a clinical trials apparatus.

Second, the startup of Genomed occurs at a time when the Indian pharmaceutical industry is in the process of retooling and reimagining its business models. Traditionally, the Indian industry drew its strength from reverse engineering generic versions of drugs that were already on the market. This was made possible by the Indian Patent Act of 1970, which allowed only process, and not product, patents on drugs, thereby allowing a patent on the method of manufacturing a drug, not on the drug molecule itself. In 1995, however, India became signatory to the agreement on Trade-Related Aspects of Intellectual Property Rights (TRIPS) of the World Trade Organization (WTO), which mandated a product patent regime. India was given a ten-year grace period to comply with the new patent regime. Knowing that relying on generic drugs to sustain their businesses was no longer going to be tenable, Indian pharmaceutical companies needed to look for other sources of revenue. Some of the larger Indian companies decided to invest in research and development (R&D) capabilities so that they could develop novel drugs in the manner of large Western companies.

Third, Wellspring must be located in the context of the early stages of building infrastructure for clinical trials in India.[8] Clinical trials became important for India on two counts. In an R&D-based environment that involves the development of novel therapeutics, clinical trials are required to establish drug safety and efficacy. Meanwhile, as a service provider to other companies, conducting clinical trials could become a revenue source in and of itself, especially at a time when United States–based companies were looking to outsource clinical trials in the late 1990s for reasons of cost and ease of patient recruitment.

Wellspring, therefore, is in its very imagination more an experimental site rather than a therapeutic one. It is, indeed, a "five-star" hospital in appearance: glittering marble floors, comfortable sofas littered around the hallways, and hospital beds with bright yellow bedcovers all make Wellspring Hospital seem more like a hotel than a hospital. What makes Wellspring even more unlike anything resembling "normal" Indian hospitals is the striking and almost complete absence of patients. What provides a fourth, political economic context for Wellspring is the larger urban ecology within which it is situated.

Wellspring is located in Parel, the heart of downtown Bombay, but also the heart of the part of Bombay that houses the textile industry. Bombay's economy largely grew on the strength of a textile industry that rapidly disintegrated through the 1980s and 1990s, leaving visible from the windows of Wellspring the empty shells of once prosperous mills. The population of Parel consists largely of unemployed mill workers who have gone through periodic cycles of unionization over the previous decade, but whose struggles to reoccupy and reopen the failing mills ended in defeat.[9] A hospital like Wellspring therefore abuts both the poverty of recent deindustrialization and the new wealth that displayed through other monstrously glamorous erections, such as nearby shopping malls that sell foreign brand-name consumables that can be afforded only by the rapidly ascendant middle class.

Wellspring's decision to locate in Parel was almost certainly not accidental. Available to company researchers was a huge unemployed local population that ended up being easily recruited into clinical trials, which do after all compensate volunteers. The ethics of such trials can be understood and evaluated only within the local ecologies in which they are conducted, ecologies that problematize the very notion of a trial "volunteer" in such a context. Just as Marx describes the forced proletarianization of the working class during the Industrial Revolution in volume 1 of *Capital* (Marx [1867] 1976, 873–942), so one can see how forced deproletarianization as a consequence of the virtual death of an industry in Bombay leads in Parel to the creation of a new subject population created as targets of experimental therapeutic intervention (Sunder Rajan 2007a). This dynamic endures even when the particularities of the situations change. Hence, though Parel itself has become gentrified, I learned in an interview with the chief executive of another Bombay-based CRO that a major site for recruiting subjects into clinical trials is now the Western port city of Surat, which has seen the disintegration of its diamond industry. Apparently, it is the unemployed diamond workers who become the primary experimental subjects in these trials.

What is at stake here is not simply a judgment on the ethics of clinical trial recruitment strategies on their own terms, but rather the question of how pararegulatory and paraethical regimes of pharmaceutical governance come into being. Specifically, at stake is an understanding of the relationship between national and global enterprises of clinical trials and local forms of indebtedness.[10] In this process, biosociality itself gets configured as a relationship between vendors and clients.

The pharmacogenomics work at Wellspring-Genomed was explicitly strategized as research that could be of interest to Western biotech or pharmaceutical companies that might wish to outsource clinical trials.[11] The resource that makes this attractive is not only India's emergent pharmacogenomic capabilities but India's *population*. As CBT Director and Genomed board member of, S. K. Brahmachari, pointed out in a personal interview, India's cross-section of populations covers the spectrum of the world's populations. Therefore, "if they want Caucasians, we'll give them Caucasians; if they want Negroids, we'll give them Negroids; if they want Mongoloids, we'll give them Mongoloids."[12] In this view, Parel becomes a very melting pot of clinical trials, and the Indian state acts (through the company it seeds) as a full-blown market agent making Indian populations available as experimental subjects to Western corporate interests. In the process, a genome startup that was seeded as part of India's genome initiative came to be involved in an enterprise centered primarily on clinical trials.

These experimental subjects are not sovereign consumers, as Genomic Health's potential consumers are. But they are inserted in the same global market logic, and indeed in the emergent technological and epistemic logics of genomics. Instead of being consumers, they are *consumables*—their bodies made targets of experimental therapeutic intervention so that other bodies in other locations can benefit from drugs as sovereign consumers. Meanwhile, the sovereign contracting agents here are the stakeholders in Wellspring-NPIL and the Indian state. The state, moreover, delineates the conditions under which global clinical trials can be conducted in India, through the construction of regulatory laws and guidelines. This phenomenon in India goes beyond the specific case of Wellspring-Genomed; it is attested to by the huge level of capacity building for global clinical trials going on in the country, especially since 2005.[13]

Experimental subjects, of course, exist also in the United States, and historically, these subjects, especially for early-stage clinical trials, have tended to belong to marginal, excluded, or otherwise dispossessed populations. In the 1970s, as clinical trials became more and more important to drug development, the primary pool of experimental subjects came from the prison population. The use of prisoners as experimental subjects in clinical trials was prohibited by law in 1981.[14] The situation in Parel is slightly different, having to do not with the incarcerating role of the state, but with a political economy of dispossession that although regulated in crucial ways by the state, has to do with global capital flows and changing local structures of capital (Sunder Rajan 2007b).

In spite of historic and ongoing inequities in access to health care in the United States, clinical experimentation still operates under an implicit social contract that assumes that some people get experimented upon in clinical trials so that drugs will be available to the public writ large—even though the price of drugs in the United States makes accessibility a serious concern. India, on the other hand, does not yet constitute a sizeable market for novel drugs that come out of clinical trials, in part because of lower purchasing power, which makes India a less profitable drug market from the perspective of multinational pharmaceutical companies, and in part because until India became a signatory to the WTO, the local market for drugs was largely in generics, which do not require an elaborate clinical testing regimen prior to their manufacture. In other words, experimental subjectivity, in the Indian context, is much less closely coupled to the social contract of therapeutic consumption than in the United States. Subjects such as those in Parel are therefore merely risked without the attendant benefits.[15]

Without creating an essentialist binary between the United States and India with regard to experimental subjectivity, I do wish to suggest that with the Wellspring case is emblematic of India's structural position within global regimes of biocapital, a position in which the individual is not so much a sovereign subject as *subjected to* the circuits of clinical experimentation. It is impossible to understand individual interests in Parel in terms of consumer preference or personalized choice—the vocabulary that to a Randy Scott is entirely commonplace. (Of course, even the "sovereign" subject in the U.S. case is configured as always already a consumer, and in that sense is not entirely a free agent.)

Subjectivity and/as Subject-Constitution

In this chapter, I am thinking through the question of subjectivity as it gets constituted in emergent biomedical regimes globally. I do so not by talking to subjects themselves, either patient-consumers in the United States or experimental subjects in India, though that is undeniably an important project. Instead, I am interested in exploring the dynamics of global capital as it interplays with the retextualizing of the postgenomic human subject.[16]

So why focus on subjectivity at all? I suggested in the introduction to this essay that subjectivity can refer to both subjection and subjectivation: in other words, both to the ways in which individuals are subjected to power and as they gain power as autonomous actors with agency. Through the cases presented here, we can in fact discern three strands of what might be called "subjectivity":

• The first concerns how people who are subjected to new biomedical technologies—whether as therapeutic consumers or as experimental subjects—might think about their enrollment and experience in this enterprise. Those self-understandings, of course, are never purely a function of individual agency, but get shaped by a number of social and cultural factors. For instance, Mary-Jo Good (2001) writes about the narratives of hope that are constructed by doctors with their patients, and Rayna Rapp (2000) shows how women's responses to amniocentesis are mediated by factors such as class and race. In short, the biomedical regime within which a subject is located does not predetermine all subjective responses and meanings.

• The second strand concerns the ways in which subjectivity materializes through the varied constructions of biomedicine, bioethics, regimes of value and capital, and—most important for our purposes—law. These renditions of the subject interact with subjectivity in the first sense largely through people's understandings of the classifications and categories made up by these powerful professions, as well as of capitalist divisions of wealth and labor. These are subject positions that people are able to learn and negotiate, however unequally or unevenly, and sometimes manipulate.

• The third strand has to do with the statistical categories that get constructed, either by clinical trial methodologies, or by population genetics, or by forecasting metrics based on bioinformatics that predict therapeutic markets, and thereby provide an epistemic basis for the imagination of a patient as a consumer in a marketplace constructed around "surplus health." This is a subjectivity that has more to do with the textuality of

DNA, and the biopolitics of genomically constituted populations more generally, than with individual understandings or agency.

The second and third strands are examples of what I earlier referred to as *subject-constitution*. The second strand refers primarily to the institutional aspects of subject-constitution and the third refers primarily to its epistemological aspects. Of course, the institutional and epistemological aspects are continually coproduced; they are never easily separable from each other, and certainly neither is an a priori determinant of the other (Jasanoff 2004).

My treatment of subjectivity in this essay focuses on the second and third strands and is thus different from the analyses of subjectivity that one sees in medical anthropology—as exemplified in the work of Good or Rapp. Nor does my account resemble the delineation of postcolonial pharmaceutical economies, perhaps most powerfully illustrated by Joao Biehl's narration of the life story of Catarina, an inmate of Vita, a "zone of social abandonment" in Brazil where people are simply left to die (Biehl 2005). My subjects are also different from the sort of agential experimental subject that Nathan Greenslit describes in his work on the marketing and therapeutic consumption of antidepressants; in his account, people who use drugs develop their sense of self in relation to what the drugs do for them, and are therefore "experimental subjects" because they experiment on themselves (2007). This is neither to trivialize nor critique such strands of work—quite the contrary—but rather to assert that when "subjectivity" is used in the STS or medical anthropology literatures, it quite possibly refers to substantially different things across different bodies of work.

My attempt here, then, is to elaborate on the joint institutional and epistemological construction of biomedical subjectivity, which involves an abstracted imagination of people away from their real conditions of life or existence. It involves the imagination and construction of an anxiety-ridden, or perhaps desperately hopeful, person taking a genetic diagnostic test as a consumer who can be persuaded to buy a product in order to grow a market. It involves the imagination and construction of an unemployed textile or diamond worker, wondering how to provide for his family, as a statistical data point in a clinical trial that can generate knowledge that travels seamlessly across bodies and geographical boundaries. My anthropological work is to precisely study these abstractions that eventually find their way back into bodies, and one way to do this is to juxtapose incongruent modes of subject-constitution in different locales next to one another.

The stakes in making the juxtaposition, however, are not simply comparative. Wellspring and Genomic Health are not two distinct case studies that operate independently of each other. Although there is no tangible relationship to speak of between these two particular entities—certainly there was not one at the time (early 2000s) when I was researching them—structurally, the stories that I tell are interlinked. This is because it is the health of the therapeutic consumer (wherever she might be located) in whose cause the experimental subject gets experimented upon in the first place. That this health too gets abstracted and made valuable as a source of surplus is important. The experimentation of the Parel clinical trial occurs in the service of a therapeutic consumer who is herself alienated from her healthiness, precisely because her very construction as a patient is predicated on her being a consumer of health. In other words, the intuitive idea of health, as having to do with an individual's well-being (and as something that, in an ideal welfare state, might even be considered a public good), has itself been transformed.

What I am tracing, therefore, is not just the differential constitution of biomedical subjectivity in two distinct national locales, but as well the differential constitution of biomedical subjectivity in globally interconnected regimes of knowledge and economics, at a historical moment when health comes to be appropriated by capital. This appropriation of health by capital is not simply an institutional story of biomedical research being performed in corporations, with corporate agendas driven by motives of private profit, though that is unavoidably a part of the story. It is rather a more deeply ideological and epistemological account, speaking to the conditions of possibility whereby it even becomes possible to imagine people as biosocial consumers and bioavailable consumables, without explicitly reflecting on the silenced interrelationships, appropriations, and dispossessions that are required for such an imagination to be actualized into reality. This modality of subject-constitution forms the substance of the intertwined histories of industrial and postindustrial capitalism and political economy—and now the life sciences emerge as foundational epistemologies that underwrite that same dynamic.[17]

What unites the therapeutic consumer in the Genomic Health story and the experimental subject in Parel, then, is that *both types of subjectivity are always already available to be appropriated by capital.* I am not making a determinist point here. I am neither saying that these are the only sorts of subjectivity that are possible nor that resistance is impossible or futile. My aim is to show how it becomes possible for these two kinds of subjects, so separate geographically and in terms of class position, to become so easily, naturally, and seamlessly imaginable as *valuable*—and as valuable in particular ways—to capital. This imagination is simultaneously institutional (brought about by biomedical knowledge operating in the context of corporate value generation) and epistemological (genomics as constructing individuals who are predisposed to disease or showing differential responses to drugs between individuals and populations that could become commercially valuable).

Bioconstitutionalism and the Genomic Subject

The constitution of the genomic subject by entities such as Genomic Health or Wellspring-Genomed represents a biomedical point of view—a form of situated knowledge or situated rationality (Haraway 1991)—that makes sense if the life sciences are involved in an enterprise of value generation. Interestingly, of course, this speculative capitalist viewpoint can just as easily be adopted by the Indian nation-state as it can by a Silicon Valley entrepreneur, though the specificities of the two viewpoints depend upon the institutional and legal contexts within which they take shape. This biomedical viewpoint is not to be confused with the views of the people who are actually subject to these corporate and state-corporatist interests. My point is that, in this viewpoint, the perspectives of the latter are dismissed as irrelevant, even as nonknowledge, at the very moment when people are valorized as agential, biosocial patient-consumers. We see this dismissiveness when Randy Scott says that the sovereign consumers in whose cause a company like Genomic Health comes into being need to consume its products because they don't know the diseases coming at them in the future. This is a statement about the unreliability of everyday epistemologies (we might think we are normal, but really we are not). It is a statement that, at the very moment at which it reconfigures individuals as always already patients-in-waiting, who are in turn consumers-in-waiting, does so through assertions of fact that by definition are unknowable by the very subject who is configured as consumer. Such a subject might well have agency, but it is an agency that in the construction of the health status of the consumer-in-waiting is rendered epistemologically null. The biomedical viewpoint on the health of the consumer-subject is legitimized as the only epistemologically valid one.

The experimental subject, too, is similarly constructed by biomedicine—by the statistical imagination of evidence-based medicine that renders clinical trials necessary (for which, see Daemmrich 2004) and by legal and political economic factors that make subject recruitment in the United States both difficult and expensive, thereby making outsourcing attractive. If the sovereign consumer subject is epistemologically silenced, then the Third World experimental subject is ethically silenced, retaining little in the way of moral agency. Building capacity for clinical trials in India is based on the establishment of an elaborate regulatory infrastructure that insists upon good clinical practice (GCP), focusing on the collection of informed consent, and the protection of the privacy and anonymity of these subjects (Sunder Rajan 2007a). This elaborate insistence on the forms of ethics constitutes these subjects as volunteers in clinical trials. In the imaginary of the state's biomedical and regulatory apparatus, these are agential, autonomous, and rational subjects who freely consent to participate in clinical trials. It is not just the histories of dispossession through disintegrating urban industries that are effaced in this construction of the "free" experimental subject. It also makes invisible the person of the experimental subject. Indeed, my own ethnographic knowledge of the class position of the experimental subjects at Wellspring or in Surat comes from scientists or managers of the CROs; I cannot get ethnographic access to these subjects precisely because of the ethical apparatus that anonymizes and so "protects" them. Even the mill workers' unions in Bombay, which were actively involved in representing the workers in wage and tenancy-related issues, were unaware that some of these very workers were being recruited into clinical trials. If the sovereign consumer is rendered unknowing of his or her own bodily prospects at the same time as he or she is granted agency to consume health, then the experimental subject is rendered unknowable at the same time as he or she is granted agency to "voluntarily" submit to clinical trials.[18]

Both the consumer subject and the experimental subject are examples of biomedical subjects who are (literally) incorporated into a global apparatus of value generation that materializes differentially in different locales at particular moments in time. Such differential materialization is hardly predetermined, but it is also, as I have tried to show, not merely contingent. The recent celebration of contingency in the anthropology of science is in many ways well founded (Rabinow 1999; Ong and Collier 2004), but that move provides insufficient explanation for the ways in which these differential materializations get ordered along—and indeed help reinscribe—the lines of historical inequities. In this case, one sees the reinscription of both class and postcolonial inequities in the playing out of contemporary global biomedicine. A lively question this opens up for historical inquiry concerns the relationship between capitalist and colonial constructions of subjectivity on the one hand and biomedical ones on the other. We see in this case how modern biomedicine reactivates historical relations of wage labor as relations now of experimental subjectivity, just as it reactivates transnational relations in ways that resonate with older colonial logics that involved the consumption of Third World resources for First World gain.

Understanding the nature of such appropriation involves situating an entrepreneur such as Randy Scott (who, on his own, might seem to be articulating "bottom-up" biosociality of the sort argued for by Rabinow and Rose) next to the creation of experimental subjects in Parel. In his afterword to a recent volume on biosociality, Rabinow refers to the linking of different "domains of practice" by the concept. These include, in his articulation, "genomics, pastoral care, patients' rights, alliance with researchers, etc." (Rabinow 2007a). The situation of experimental subjectivity that I narrate figures as part of that elided "etc." But it is the constitutively necessary "etc." required for this particular form of speculative, capital-intensive global biomedicine to function. And it feeds into an enterprise that even in the United States is not, from the perspective of the pharmaceutical industry, about pastoral care—except as incidental to the generation of market value.

This is why the question of subjectivity, to my mind, is directly related to the question of "reframing rights" and the concerns of bioconstitutionalism that animate this volume. Subjectivity, especially if one considers the diverse modalities of subject-constitution as I have sketched, provides an empirical and conceptual lens onto the naturalized, taken for granted, and hidden ways in which regimes of knowledge and value are in fact constructed in extremely complicated ways. In this essay, I have offered a glimpse into the striations, the differentiations, the articulations, the naturalizations—and indeed the elisions and the silences—that structure the operation of contemporary biocapital in an emerging global constitutional order.

Notes

1. For a further elaboration of the relationship between a Marxian attentiveness to modes of production and Sheila Jasanoff's insistence on coproduction as an analytic framework through which to study contemporary technoscience in sociopolitical context, see my introduction to the forthcoming *Lively Capital* (Sunder Rajan, forthcoming).

2. This dual conception of subjectivity is developed, for instance, in Hegel's discussion of consciousness, desire and recognition in *The Phenomenology of Spirit*

(Hegel [1807] 1977). It animates the idea of the revolutionary subject in Marx and Marxism, just as it does the idea of the subject in Freud and psychoanalysis. Even philosophers like Foucault, who attempt to distance themselves from Hegelian dialectics, cannot escape this ambivalent conception of the subject, as simultaneously disciplined and agential/productive.

3. This is stated on Genomic Health's webpage, <http://www.genomichealth.com/message.htm> (accessed July 2004).

4. Randy Scott, talk given at the Genome Tri-Conference, San Francisco, March 7, 2001.

5. Randy Scott, talk given at the Genome Tri-Conference, San Francisco, March 7, 2001.

6. "Surplus health" refers to the potential for therapeutic consumption over and above that required to maintain healthiness in the conventional sense of freedom from disease. In the process of generating this surplus, health becomes abstracted and divorced from concerns with healthiness while also getting appropriated by capital (Dumit, forthcoming). For an argument about how surplus health logics in the United States exacerbate the situation of experimental subjectivity in India, see Sunder Rajan 2007a. Dumit and Sunder Rajan (2007) attempt to bring these arguments together in order to suggest that consumer genomics has entered a postbiopolitical phase. On this point, we differ from Rose (2006).

7. For a detailed account of CBT, see Sunder Rajan 2006. CBT has now been renamed the Institute for Genomics and Integrative Biology (IGIB), and the person who headed CBT at the time I did my research there, Samir Brahmachari, is now head of India's Council for Scientific and Industrial Research (CSIR).

8. Since then, capacity building for clinical trials in India has grown at a rapid rate. See Sunder Rajan 2007a and 2010 for an account of this.

9. Indeed, in the few years since I initially wrote this account, Parel has gentrified beyond recognition. Many of the unemployed workers have been displaced, mostly to the urban peripheries of Bombay. There is still some union and civic activism for workers' rights, but the demographics of the locality have changed significantly.

10. The relationship of bodies as sites of medical intervention to local forms of indebtedness is beautifully illustrated in Lawrence Cohen's (1999) account of organ transplantation in South India.

11. It should be pointed out that at the time this account was researched—and still, at the time of writing—it is illegal to conduct global, first-in-country, Phase I trials on healthy volunteers in India. This situation is almost certain to change in the near future, as CROs have been lobbying regulatory agencies to rule otherwise. While infrastructure was being built to conduct outsourced clinical trials from abroad, only late-stage clinical trials could be performed. Any Phase I trials conducted at that time would have had to be for Indian sponsors.

12. S. K. Brahmachari, interview with the author, January 7, 2002. The difficulty of classifying populations for population genetics is constitutive to its epistemology. See Reardon 2001 and 2004 for an account of the difficulties encountered by the Human Genome Diversity Project as a consequence of this fact.

13. See Sunder Rajan 2007a for an account of this.

14. For an account of experimental subjectivity in clinical trials in the United States, and the way in which clinical trials must be located in the context of the neoliberalization of healthcare in this country, see Fisher 2009. Adriana Petryna makes the argument that the "loss" of prisoners as a potential experimental population has been a major factor in increasing the global outsourcing of clinical trials from the United States to other countries (2009). For a much broader historical account of clinical experimentation in the United States through the twentieth century, see Marks 1997.

15. India's status as a drug market is likely to change over the next few years. For a study of the marketing of psychopharmaceuticals in India by multinational companies, see Ecks 2010. At the moment, however, the major drug markets in the world are constituted by the United States, Europe, and Japan, which together constitute approximately 85 percent of the market for patented medications.

16. This, I believe, is entirely consistent with the themes of this volume, even if a number of other contributions focus most explicitly on national or global governance rather than on capital. It should be understood that capital and the nation-state have been mutually implicated through the history of liberalism and capitalism, and the nature of that implication needs fresh empirical interrogation. The life sciences, law, and their coproduction provide a rich site such where such interrogation can occur.

17. Dominic Boyer has usefully reminded me that Adam Smith's famous example of a pin factory is precisely such an exercise in the abstraction and naturalization of violent histories as natural laws. To quote Boyer: "Adam Smith's pin factory was a good example, one in which the estrangement generated by an advanced division and specialization of labor, further mediated by a monetary economy, monetarized wage labor, and institutionalized private property was suppressed under the assertion that it was in human nature to 'truck, barter and exchange' and that the productivity of capital worked to the general benefit and prosperity of mankind. Marx's enduring contribution to social theory of knowledge was to show how the conditions of Smith's lifeworld (especially the universalization of wage labor) generated Smith's impression of its 'natural' status as well as the possibility of a general, transhistorical category of 'Labor' and its productive counterpart, 'Value'" (Boyer 2008).

18. I use "him" to describe the experimental subject because experimental subjects in clinical trials in India are typically male, unless the trial sponsor specifically calls for female subjects in the trial protocol.

References

Biehl, Joao. 2005. *Vita: Life in a Zone of Social Abandonment*. Berkeley: University of California Press.

Boyer, Dominic. 2008. Personal correspondence. September 5.

Cohen, Lawrence. 1999. Where It Hurts: Indian Material for an Ethics of Organ Transplantation. *Deadalus: Bioethics and Beyond* 128 (4): 135–164.

Collins, Francis, and Victor McCusick. 2001. Implications of the Human Genome Project for Medical Science. *Journal of the American Medical Association* 285 (5): 540–544.

Daemmrich, Arthur. 2004. *Pharmacopolitics: Drug Regulation in the United States and Germany.* Chapel Hill: University of North Carolina Press.

Dumit, Joseph. Forthcoming. *Drugs for Life: Managing Health and Happiness through Facts and Pharmaceuticals.* Durham, N.C.: Duke University Press.

Dumit, Joseph, and Kaushik Sunder Rajan. 2007. "Biocapital, Surplus Health and Clinical Trials: Toward a Health Theory of Value." Paper presented at Experimental Systems conference, University of California, Irvine.

Ecks, Stefan. 2010. Global Citizenship Inc.: Big Pharma and Depression Awareness in Urban India. In *Asian Biotech: Ethics and Communities of Fate* ed. Aihwa Ong and Nancy Chen, 144–166. Durham, N.C.: Duke University Press.

Fisher, Jill. 2009. *Medical Research for Hire: The Political Economy of Pharmaceutical Clinical Trials.* New Brunswick, N.J.: Rutgers University Press.

Foucault, Michel. 1980. *Power/Knowledge: Selected Interviews and Other Writings, 1972–1977.* New York: Pantheon.

Good, Mary-Jo. 2001. The Biotechnical Embrace. *Culture, Medicine, and Psychiatry* 25 (4): 395–410.

Greenslit, Nathan. 2007. *Pharmaceutical Relationships: Intersections of Illness, Fantasy and Capital in the Age of Direct-to-Consumer Marketing.* PhD diss., History and Social Studies of Science and Technology, Massachusetts Institute of Technology.

Haraway, Donna. 1991. *Simians, Cyborgs, and Women: The Reinvention of Nature.* New York: Routledge.

Hegel, Georg. [1807] 1977. *Phenomenology of Spirit*, trans. A.V. Miller. Oxford: Oxford University Press.

Jasanoff, Sheila. 2004. Ordering Knowledge, Ordering Society. In *States of Knowledge: The Co-Production of Science and Social Order*, ed. Sheila Jasanoff, 13–45. London: Routledge.

Marks, Harry M. 1997. *The Progress of Experiment: Science and Therapeutic Reform in the United States, 1900–1990.* Cambridge: Cambridge University Press.

Marx, Karl. [1867] 1976. *Capital: A Critique of Political Economy, Volume 1,* trans. Ben Fowkes. London: Penguin Books.

Novas, Carlos, and Nikolas Rose. 2004. Biological Citizenship. In *Global Assemblages: Technology, Politics, and Ethics as Anthropological Problems*, ed. Aihwa Ong and Stephen Collier, 439–463. Malden: Blackwell.

O'Malley, Patrick. 2000. Uncertain Subjects: Risks, Liberalism and Contract. *Economy and Society* 29 (4): 460–484.

Ong, Aihwa, and Stephen Collier, eds. 2004. *Global Assemblages: Technology, Politics, and Ethics as Anthropological Problems.* Oxford: Blackwell Publishing.

Petryna, Adriana. 2002. *Life Exposed: Biological Citizens after Chernobyl.* Princeton: Princeton University Press.

Petryna, Adriana. 2009. *When Experiments Travel: Clinical Trials and the Global Search for Human Subjects*. Princeton: Princeton University Press.

Rabinow, Paul. 1992. Artificiality and Enlightenment: From Sociobiology to Biosociality. In *Incorporations*, ed. J. Crary and S. Kwinter, 234–252. New York: Zone Books.

Rabinow, Paul. 1999. *French DNA: Trouble in Purgatory*. Chicago: University of Chicago Press.

Rabinow, Paul. 2003. *Anthropos Today: Reflections on Modern Equipment*. Princeton: Princeton University Press.

Rabinow, Paul. 2007a. Afterword: Concept Work. In *Biosocialities, Genetics, and the Social Sciences: Making Biological Identities*, ed. Sahra Gibbon and Carlos Novas, 188–192. London: Routledge.

Rabinow, Paul. 2007b. Marking Time: On the Anthropology of the Contemporary. Retrieved from <http://anthropos-lab.net/documents/pubs/marking-time-on-the-anthropology-of-the-contemporary/> (accessed December 19, 2010).

Rapp, Rayna. 2000. *Testing Women, Testing the Fetus: The Social Impact of Amniocentesis in America*. New York: Routledge.

Reardon, Jenny. 2001. The Human Genome Diversity Project: A Case Study in Coproduction. *Social Studies of Science* 31:357–388.

Reardon, Jenny. 2004. *Race to the Finish: Identity and Governance in an Age of Genomics*. Princeton: Princeton University Press.

Rose, Nikolas. 2006. *The Politics of Life Itself: Biomedicine, Subjectivity, and Power in the Twenty-First Century*. Princeton: Princeton University Press.

Sunder Rajan, Kaushik. 2005. Subjects of Speculation: Emergent Life Sciences and Market Logics in the U.S. and India. *American Anthropologist* 107 (1): 19–30.

Sunder Rajan, Kaushik. 2006. *Biocapital: The Constitution of Post-Genomic Life*. Durham: Duke University Press.

Sunder Rajan, Kaushik. 2007a. Experimental Values: Indian Clinical Trials and Surplus Health. *New Left Review* 45:67–88.

Sunder Rajan, Kaushik. 2007b. Biocapital as an Emergent Form of Life: Speculations on the Figure of the Experimental Subject. In *Biosocialities, Genetics, and the Social Sciences: Making Biological Identities*, ed. Sahra Gibbon and Carlos Novas, 157–187. London: Routledge.

Sunder Rajan, Kaushik. 2010. The Experimental Machinery of Global Clinical Trials: Case Studies from India. In *Asian Biotech: Ethics and Communities of Fate*, ed. Aihwa Ong and Nancy Chen, 55–80. Durham, N.C.: Duke University Press.

Sunder Rajan, Kaushik. Forthcoming. *Lively Capital: Biotechnologies, Ethics and Governance in Global Markets*. Durham, N.C.: Duke University Press.

10

Human Population Genomics and the Dilemma of Difference

Jenny Reardon

The first decade of the twenty-first century witnessed significant institutional changes in the governance of human population genomics. Gone are the days when, for a few bottles of medicine, or some salt and beads, a population geneticist could go to the "remote corners" of the earth to sample human genetic material thought to contain the secrets of human evolution (Cavalli-Sforza, Menozzi, and Piazzi 1994). Today, those who traverse the globe with syringes, dry ice, and Eppendorf tubes to collect human DNA also need lengthy informed consent forms, protocols for community engagements, and a relatively new, yet necessary, class of expert professionals—bioethicists. It is hoped that traveling with such an elaborate ethical apparatus will enable genome scientists to bring a Western liberal democratic system of rights to the diverse geographic locations in which they sample—rights which today are often as critical as blood itself to the functioning of global genomic information systems.

In addition to extending the geographic scope of these rights, administrators of large public genome sampling efforts have sought to expand the kinds of subjects who might lay claim to them. Responding to changes in the nature of genetic research—from mapping an individual human genome to studying human groups for variation among genomes—these administrators have sought to forge novel ethical practices that would expand a Western framework of rights from individuals to groups (North American Regional Committee, Human Genome Diversity Project 1997). Most notably, in 1995 organizers of the Human Genome Diversity Project—an initiative that sought a global survey of human genetic diversity—proposed the novel ethical practice of *group consent*. Proposals for group review, community consultation, and community engagement soon followed (Sharp and Foster 2000). All of these developments sought to accommodate the shift in focus in human genomics from studying individuals to studying groups. In particular, they responded to historically embedded concerns about the consequences of objectification. Rather than stripping human beings of their individuality and rights by rendering them objects of scientific analysis, organizers of genomic studies of human differences envisioned the opposite: the production of human subjects empowered with new rights.

Although designed and launched with great hopes, and even a claim of reaching a new ethical "gold standard" (King 1996), a decade into efforts to implement these new rights found genome scientists, their subjects, and associated bioethicists and social scientists concerned about their further institutionalization.[1] In this chapter, I document and explain these concerns through an analysis of two moments in which the offer of these new rights generated unease: first, the creation of and response to the Human Genome Diversity Project's group consent provision; second, the National Institute of Health (NIH) decision to require community engagement as part of collecting samples for the International Haplotype Map Project (HapMap). I argue that both attempts to craft new rights of self-determination for groups faltered as they enabled scientists to elide responsibility for the orderings and valuations of human differences that undergirded their offers of new rights.

The chapter seeks to demonstrate how this uncomfortable trade-off happened through documenting how genomics administrators' conception and enactment of rights were formed in tandem with their conception of human population genomics. In particular, I demonstrate how this process of coproduction proceeded in a manner that rendered invisible the role that institutions (in this case, institutions of law and science) played in constructing the groups that would be granted the new rights for designing and reviewing genomic research (on coproduction, see Jasanoff 2004). In the case of the Diversity Project, organizers bypassed responsibility for questions about how human differences should be ordered into groups, and by whom, by taking these differences as given by nature, and thus discernable by human population genetics. In the wake of the controversies sparked by the Diversity Project, this "ethical bypass" gained much attention.[2] The puzzle I address in this chapter is how subsequent human genome variation research projects continue to bypass responsibility for their roles in co-constituting natural and moral orderings of human difference, despite efforts to address ethics at the earliest stages of research design. Comparison of the Diversity Project and HapMap cases proves illustrative.

Organizers of the HapMap explicitly attempted to respond to the critiques sparked by Diversity Project organizers' positioning of their "science" as the arbitrator of politically and socially consequential decisions about the order of human differences. However, HapMap organizers' own efforts to resolve questions about the nature and proper order of human differences faced similar difficulties. Although intent on drawing distinct lines between their initiative and the Diversity Project, HapMap organizers' fundamental assumptions remained the same. Like the drafters of the Diversity Project's group consent provision, they too believed that human difference exists a priori, and that the right expertise can know and characterize that difference. The HapMap designers differed only in where they located the source of expertise: whereas the Diversity Project organizers found this expertise in science (in the form of human population genetics), the administrators of the HapMap found it in both science and people (in the form of "communities"). In neither case did an occasion arise to reflect on the deeper question that population genetic studies of human differences poses: when and how do genetic studies of human differences classify human beings into groups in a manner that can lead to stigmatization, and when and how do these acts of ordering illuminate and promote desirable social ends?[3] Instead, project organizers found themselves in effect in a bioconstitutional moment posing fundamental questions about how human differences should be recognized and valued, yet working within conceptual structures that failed to address these questions.

In particular, in casting ethics as distinct from science—as something that does not inhere in science but instead needs to be done along with it—Diversity Project and HapMap organizers both missed the ways in which their "expert" understandings of human differences themselves entailed decisions about which bodies and which moral commitments state institutions might recognize. Further, the liberal democratic approach to rights that formed the basis of the official "ethics" arms of both the Diversity Project and the HapMap reinforced this elision through positing the innate existence of human differences. Thus, as I illustrate in this chapter, efforts to conduct human population genetics responsibly while extending Western liberal democratic rights from individuals to groups failed to account for and respond to the role that political ideas and practices play in constructing the sciences of human difference—and, by extension, seemingly neutral science-based ethics. As a result, rights became severed from response-ability (Haraway 2008, 88).

Not surprisingly, trouble resulted. Although observers initially celebrated the effort by both Diversity Project and the HapMap organizers to forge new rights along with new forms of science (The International

HapMap Consortium 2004), failure to adequately address questions about the appropriate basis for group designations disturbed scientists and non-scientists alike. In particular, as discussed herein, population geneticists and their potential research subjects objected to novel ethical practices that granted rights to groups without providing a mechanism for considering whether the newly empowered groups had any legitimate natural or social existence. Such practices, they argued, bypassed existing sources of authority for deciding how and when human differences should be recognized and used to define populations.

The chapter ends with a reflection on what it might take to reframe rights and population genomics in a manner that entails taking responsibility for decisions about how human differences gain recognition, value, and form. What is needed is neither a rights discourse that passes off responsibility for defining and valuing human differences to population genomics nor a population genomics that displaces responsibility for these critical questions onto a rights discourse. Instead, rights must be rejoined with responsibilities.

The Emergence of the "Population" for Biology and Rights

In the late 1970s, the political theorist Michel Foucault described modernity as an era marked by the emergence of "the population" as an object of knowledge and governance (Foucault 1976). Through what Foucault labeled "the biopolitics of the population," birth and death rates, levels of health, life expectancy and longevity all emerged as sites of intervention and supervision—sites at which life itself became an object and instrument of power (139). Thirty years later, Foucault's provocative claim would be hard to deny. Despite official rhetoric that stresses the biological similarity of all humans (Hotz 1995; Venter 2000), institutions of science policy, in alliance with the life sciences, now invest tremendous resources in characterizing human biological differences at the population level (Duster 2003; Reardon 2005; Steinberg 1998).

The move to constitute the "population" as a central object of biomedical research gained renewed support in the United States during the health disparity debates of the 1980s, prior to the genomic age. At that time, leaders of research institutions and health policy makers began to confront complaints from some politicians, biomedical researchers, and health activists who argued that the biomedical system was exclusionary and inattentive to social differences that matter (Epstein 2004). These critics argued that in practice biomedical research institutions supported clinical trials only on white heterosexual men, even though the results supposedly applied to all humans. Advocacy groups who claimed to speak for excluded groups demanded an end to this discriminatory practice and pushed for including "special populations" with a history of being marginalized, such as women, children, and minorities.

When the interests of politically influential groups were in alignment with the state's long-standing efforts to produce knowledge about populations, the efforts to include diverse populations in state-funded biological studies succeeded. In 1993, President Clinton signed into law an NIH Revitalization Act mandating the inclusion of women and minorities as subjects in federally funded clinical research. Additionally, in the late 1990s, several offices of women's health and minority health formed within the Department of Health and Human Services. According to the sociologist of science Steven Epstein, these developments led to the creation of an "'inclusion-and-difference' policy paradigm" that sought to include underrepresented subjects in clinical trials, and to measure differences across groups in biological disease processes and responses to treatment (Epstein 2007).

Although at first this new paradigm seemed to represent a frictionless merging of the interests of the state and social movements, resistance to it emerged as attempts to implement it proceeded on the ground. As some analysts, biomedical researchers, and doctors now argue, the new inclusionary policies force researchers, activists, and policy makers alike into the difficult position of having to defend the antiracist, egalitarian potential of the same kinds of biological studies of human differences that for centuries justified social discrimination (Proctor 1988; Schiebinger 1993). In particular, some worry about the emergence of "medical racial profiling" and resulting new forms of discrimination (Schwartz 2001). To date, however, these critiques of the "inclusion-and-difference" paradigm have mainly been the preserve of biomedical researchers and academic analysts. Activists and health policy makers have largely bypassed questions about the constitution of difference, preferring instead to use claims about the biological existence of difference to ground their demands for inclusion in research (Epstein 2003).[4]

The rise of population-based genetic research followed a different trajectory. The history of genetics has been indelibly marked by blatant examples of social institutions labeling human beings as different in order to render them marginal, invisible, and even disposable (Minow 1990;

Biehl 2005). Nazi policies of exterminating entire populations on grounds of presumed biological differences led many to recoil from the very idea of drawing such distinctions among groups of human beings. Particular concern arises in cases where scientific evidence is involved in classifying people, because the authority of science gives seeming precision and validity to such distinctions and makes them less open to challenge. Population geneticists and their anthropologist colleagues intensely debated the meaning and proper role of ordering concepts such as "population" and "race" in studies of human biological diversity in the interwar period, and continued these debates through the decades after World War II (Dobzhansky 1962; Livingstone 1962). These issues reemerged in the 1990s, when human genomic research began to merge with population genetics, again entangling in unsettling ways human group categories used by states, markets, and the life sciences (Rabinow 1999b; Reardon 2001).

In this contentious context, scientists who wished to study genomic variation in human populations could not presume that human differences are naturally ordered into groups, such as "populations" or "races." Rather, they found themselves engaged in struggles to characterize and prove the existence of meaningful groups in the first place (Burchard et al. 2003; Cooper, Kaufman, and Ward 2003). As scholars of science and technology have documented, the emergence of new scientific phenomena (such as the characterization of human populations in terms of their genomes) makes the inextricable ties between science and social commitments evident and palpable (Jasanoff 2004). When the phenomena in question involve the creation of what the philosopher Ian Hacking calls "interactive kinds" (Hacking 1999), such as human groupings based on diversity, the intertwining of epistemic and normative assumptions is especially hard to deny.

Indeed, the protagonists of the HapMap explicitly recognized these ties and self-consciously attempted to implement their own visions of an ethical human genomics.[5] Accordingly, my concern here is not, as it has been in my previous work, to bring to light or further exemplify the phenomenon of coproduction. Rather, my goal here is to begin to characterize the patterns of coproduction that come into view if one compares the Diversity Project and HapMap cases. In particular, I am interested in what was brought into view and what was erased in such processes. I show how both projects constituted a natural order (a group) along with a social/moral order (a set of rights) in a manner that prevented consequential questions about how to recognize and order human differences from gaining recognition.

The Human Genome Diversity Project

In 1991, a group of population geneticists and evolutionary biologists from the United States announced their intention to sample and characterize the world's human genetic diversity by sampling human populations across the globe (Cavalli-Sforza et al. 1991). Advocates claimed that the project would elide old concerns about the use of biological studies for the production of invidious social distinctions (Reardon 2005). Molecular studies, they asserted, had revealed that all humans are 99.9 percent identical, genetically speaking, which made racial categories meaningless from the scientific point of view and thus irrelevant for their project (Cavalli-Sforza 1994). But if this were the case, some biological anthropologists began to ask, why were organizers of the Diversity Project so intent on studying that 0.1 percent of difference? Further, why did they use racially coded, colonialist population categories such as "Caucasian" and "Negrito" to define the human beings from whom they hoped to collect genetic samples (Lewin 1993; Marks 1998)?

Persistent questions such as these soon made it evident that Diversity Project organizers would not be able to bypass long-standing concerns about biological studies of human variation by claiming to be a part of a "new science" of genomics. At least in the American context, even as the century ended, biological studies of human groups remained highly suspect. The links to Nazi genocidal policies, and to the deployment of biological analysis to justify racial discrimination in the United States, were still fresh in public memory (Jensen 1969). Given these connections, it would have been surprising if a worldwide study of the genomes of human populations could have proceeded without concerns that the resulting analysis might strip human beings of their rights.[6] If these scientists truly were part of a "new science," then, many insisted, they would have to address these concerns about human rights.

It did not take long for organizers of the Diversity Project to do just this. Within two years of the first critiques, they formed an ethics committee chaired by a lawyer, and wrote a Model Ethical Protocol that featured a provision for group consent (North American Regional Committee 1997). Group consent expanded informed consent rights from individuals to collectivities. The drafters argued that, with such consent, the "populations" the project studied would not become mere objects of study, but subjects empowered with new rights.[7]

Although intended to allay concerns about the role of genomic ideas and practices in the construction of human difference, this attempt to produce novel informed consent rights for population genomics stud-ies presented new problems. These problems did not arise from an in-compatibility between human population genetics and a liberal system of rights (i.e., a so-called clash of cultures between science and law). To the contrary, both population genetics and a liberal system of rights begin from similar assumptions about the constitution and proper ordering of human differences. In the following discussion, I present these assump-tions and demonstrate how they impeded consideration of the role that both population genetics and liberal systems of rights play in producing human differences.

Coproduction and the Dilemma of Difference

As a matter of principle, liberal democratic thought does not privilege questions about difference. Responding to dissatisfaction with institution-alized forms of power (for example, the church and the monarchy) and the distinctions they created among classes of humans, liberal thinkers of the Enlightenment—such as John Locke and John Stuart Mills—argued that all individuals should be treated the same in relation to political power and should enjoy the right to govern themselves (Minow 1990, 148). Yet as historians and critical legal theorists observe, despite its rhetoric of equality and inclusion, the rights approach potentially produces a new form of exclusion. Grounding individuals' right to govern in their "ratio-nality" (an antidote to the "irrationality" of powerful despots) implies a fundamental distinction between human beings; the "rational" must be sorted from the "irrational."

These seemingly innocuous distinctions—which supposedly follow from neutral, objective criteria—prove consequential. Under monarchi-cal or religious rule, dominant institutions overtly exercised their power to draw distinctions among human beings: the outlaw from the citizen; the saved from the damned. A Western liberal democratic approach to rights, however, conceals its power to define difference through an appeal to intrinsic differences that exist among persons. As the feminist legal scholar Martha Minow explains, "The rights approach . . . retains a gen-eral presumption that differences reside in the different person rather than in norms embedded in prevailing institutions" (1990, 108). This natural-ized conception of difference makes it impossible to ask questions about the role that social norms and institutions may play in the construction of difference. Consequently, the ways in which rationality may be selectively ascribed to certain kinds of individuals (e.g., at the time of the formation of the United States, only to white, heterosexual, propertied males) rarely attracts critical attention.

Given that genomics is also based on a naturalistic conception of human differences, it should not surprise us that rights-based discourse has easily taken hold in the domain of genomics. Genome scientists attempt to distinguish themselves from their tarnished predecessors—human geneticists who believed in racial categories—partly by distancing themselves from any act of *constructing* human differences.[8] Instead, genome scientists describe themselves as *discovering* differences that already exist, encoded in variations among individual genomes. Both these scientists and advocates of a liberal democratic approach to rights express a belief in the innate existence of human differences. From this shared belief follows a shared difficulty: an inability to imagine the role that political ideas and practices play in constructing the sciences of human difference, and by extension, in the constitution of seemingly neutral, science-based ethics.[9] It is this problem that troubled the Human Genome Diversity Project's group consent provision.

Group Consent

Following high-profile cases of the abuse of human subjects in the 1960s and 1970s, many in the United States began to view human subjects research in the life sciences as prone to abuses that required a rights framework to correct them.[10] Rights were officially extended to research subjects in the Belmont Report, the definitive American statement on informed consent. This report states that researchers must obtain the informed consent of individual research subjects, who are assumed to have needs and preferences that are entitled to respect. This ethical prescription follows from the conviction that "individuals should be treated as autonomous agents," a principle that is consistent with a liberal democratic tradition of protecting and preserving individual autonomy and rights (National Commission for the Protection of Human Subjects of Biomedical and Behavioral Research 1979).[11]

Organizers of the Human Genome Diversity Project built upon this principle and extended it to the novel context of population genetics. In this context, they argued, the research subject is not an individual but a group (i.e., a *population*). The risks are different in the two contexts: individuals may be exposed to physical harm, whereas groups may be stigmatized or disempowered in various ways. Thus, the researchers reasoned, *individual* consent should be augmented by *group* consent.

It did not escape their attention that extending informed consent rights from individuals to groups would raise contentious questions about the boundaries of a group and who could speak for the group. Consistent with the ideas of innate differences that undergird both genomics and a liberal system of rights (as described previously), organizers attempted to resolve these questions by positing that the differences that define groups exist prior to efforts to know those differences; in other words, they claimed that certain groups are "natural kinds," unaffected by any contingencies surrounding their discovery or identification (Hacking 1990). Further, they assumed that expert resources existed that could identify differences: in particular, the "sampling criteria" used by Diversity Project researchers (North American Regional Committee, Human Genome Diversity Project 1997). In other words, complex questions about the boundaries of groups could be resolved by turning to biological experts with the requisite specialized knowledge and skills (e.g., population geneticists).

These assumptions—that differences existed and that experts could discern them—made it difficult for Diversity Project organizers to recognize that their beliefs might play a role in constructing human groups. Drafters assumed that they were attempting only to sample already existing groups in an ethically responsible way. That this might not be the case, however, became clear as government agencies called upon Native attorneys to review the project's ethical practices. These attorneys focused their attention on the question of who would define the "group" of group consent. Rather than see this as a neutral issue that experts could settle, they pointed to the vital political issues of sovereignty at stake. In particular, they worried about encroachment upon sovereignty rights if experts with no accountability to Native communities were allowed to designate groups for consent purposes. Cases such as *Santa Clara Pueblo v. Martinez* (1978) upheld the rights of tribes to determine their own membership (Cohen 1982; Wilkins 1997).[12] By placing the power to determine group membership in the hands of experts—such as population geneticists—the Diversity Project, Native attorneys contended, threatened to encroach upon these basic rights. Although group consent had been proposed in the proper spirit, Native critics argued that it failed to address questions of fundamental importance to a community's self-governance: how should the group be defined and by whose authority.[13]

In short, in assuming that "biological" experts could define groups, group consent sidestepped critical questions about the role that the project's research and ethical design would play in defining the very "groupness" that was under study (Rabinow 1999a). Critics argued that by failing to provide the tools needed to bring these critical questions into focus, group consent might constitute groups in a manner that would create new political and ethical problems. In particular, they were concerned that the new ethical practice might grant scientific validity to notions of race and indigeneity that simply tracked older paternalistic and colonial categories. In this way, they suggested, Diversity Project organizers' effort to be "ethical" might reinscribe the very notions of race and difference it claimed to be undermining.

The International Haplotype Map Project: Beyond Group Consent?

Many at the NIH worried that the controversy surrounding the Diversity Project would make it impossible for scientists to conduct human genetic variation research. In particular, they were sensitive to the charge that the Diversity Project, in privileging the authority of science and scientists, had ignored legitimate concerns about group identification expressed by those targeted for sampling. As a result, when the National Human Genome Research Institute proposed its own effort to sample and study human genetic variation a decade after the abandoned proposal for the Diversity Project, the organizers of this new initiative placed heavy emphasis on their intention to consult with the groups of people they hoped to sample (now called "communities"). They believed that their effort—which would become known as the International Haplotype Map Project—would avoid treating human beings like objects—a means to an end—and would instead set a higher bar for engaging with participants as active subjects. This would require that "community engagements" accompany all sampling efforts, and that these engagements would entail giving communities new rights, including the right to shape the informed consent process and the right to form an advisory group that would oversee use of the samples. Perhaps most importantly, HapMap "communities" were offered a voice in determining how their samples would be named (The International HapMap Consortium 2004).[14]

Indeed, at the beginning of the planning for the project, organizers went so far as to place the final decision for naming the source of the samples in the hands of communities. As a "Background and Overview" document distributed at a September 2002 National Human Genome Research Institute council meeting states, "The communities themselves will ultimately decide the exact description or label by which they wish samples from their population to be identified" (Community Engagement and Sample Collection for the HapMap Project 2002, 3). On the surface, this was a substantial concession to group autonomy, recognizing that affected communities have the right to determine how their collective identities will be described, even within the context of a scientific research project. However, as project organizers tried to implement this policy on the ground, hidden discrepancies emerged between researchers' and subjects' understandings of the nature of that autonomy. The desire of the people consulted in Japan to label their samples "Asian" became emblematic. As one organizer of the HapMap explained to me:

In Japan there were a lot of people who were saying they wanted the samples to be called "Asian." Which is fine, except—well, there are a couple of issues. I mean, one, you also have samples from China, so are they "not-Asian"? Or they're just "China"? You have "Asian," and then you have "Chinese"? . . . So . . . this notion of "let the community decide for themselves what to call themselves" is not quite so simple. Everybody assumes they are going to come up with a label that's scientifically defensible, which it really isn't, because Japanese people obviously don't represent all Asians. So there are *scientific* problems with that which we tried to point out.[15]

As this Project organizer's observations make clear, most administrators of the HapMap did not give up their belief in a "scientific" approach to naming samples when they announced their intention to allow "communities" to name their samples.[16]

The project documents use the word "precision" to describe the commitment to a "scientific" approach. The call for "precision" appears early on, including in a September 2002 document that defines "communities" as the ultimate arbiters of sample labels. As this document explains, "There is consensus on the need for precision and accuracy in the descriptors that are used (e.g., descriptors that accurately reflect the nature and degree of admixture in each population, that are not inappropriately broad, and that do not confuse notions of 'race' with 'ethnicity')" (Community Engagement and Sample Collection for the HapMap Project 2002). Later, organizers stressed the importance of using "precision" in the Project's "Guidelines for Referring to HapMap Populations" posted on the Web in late 2004.[17]

Although these documents call for "precision" and even provide detailed explanations for why it is necessary, they do not provide guidance on how to produce precise descriptors of populations. Further, they do not discuss how organizers might navigate potential tensions between the criteria and values embedded in project organizers' commitment to precision, and the criteria and values embedded in the project's statement that it would grant communities the power to name their samples. Instead, in official documents and interviews with me, organizers often treated "precision" as if its meaning were neutral and obvious, and equally applicable to naming efforts regardless of whether these sought to describe natural demarcations (the researchers' aim) or socially significant ones (the communities' entitlement).

As it soon became clear in practice, however, this approach to applying the principle of precision placed the project on shaky ground. Precision turned out to have no obvious or straightforward meaning. For HapMap organizers, it meant providing geographic indicators of subjects' origins on the samples. For example, they proposed labeling the samples collected in Japan "Japanese from Tokyo, Japan."[18] They believed that adding "from Tokyo, Japan" would make the difference between a vague, potentially nationalist, and political label and a precise, objective, scientific one.

What counted as clear and precise for HapMap organizers, however, appeared inaccurate from the perspective of those in Tokyo who participated in the project. As a HapMap organizer later explained to me, these participants resisted the proposed label, arguing that "they didn't all come from this place in Tokyo, they came from all these other different places, and so they shouldn't be labeled as coming from that place."[19]

In the end, the label used was not "Japanese *from* Tokyo, Japan," but "Japanese *in* Tokyo, Japan." For organizers in the United States, this label successfully brought precision to the process of naming the population sampled in Japan. As one HapMap scientist explained to me: "Precision is valuable. . . . There are all sorts of underlying assumption if I said 'Japanese.' Whereas if I said 'HapMap JPT' [Japanese in Tokyo], you know exactly what I mean."[20] However, the value and clarity some HapMap scientists attributed to "Japanese in Tokyo, Japan" would continue to elude HapMap project participants in Japan, and they accepted the label only after a "messy negotiation."[21]

As this example makes clear, geographical, place-based labels might be meaningful for human beings who have not moved around much. However, in a globalizing world, such descriptors have no obvious meaning.[22] As the HapMap scientist who asserted that I would know exactly what he meant by "Japanese in Tokyo" later reflected:

No doubt Jewish American from Brookline is probably not very different ethnically from Jewish American in Brooklyn, or Jewish American who lives near Santa Cruz. But, on the other hand, probably the amount of migration in certain countries is less, so it may be more meaningful. So coming up with any general answer is very difficult because, again, what might be very important to avoid bias in one context might be actually insufficient information in a different context.[23]

Yet even in these reflections, the first concern remains with the objective quality of the information conveyed by population descriptors (i.e., issues of bias and insufficiency) and not the sovereignty issues that had been at the heart of the Diversity Project debates, and that had motivated organizers of the HapMap to extend naming rights to those sampled. Indeed, questions about what precision might mean to different actors are not addressed in project documents. Instead, organizers of the HapMap approached the labeling process as if the meaning of precision did not itself need to be a subject of deliberation.

In the end, communities were allowed to play a role in determining the label placed on their samples, but only insofar as that role did not interfere with the values of neutrality and objectivity embedded in project organizers' vision of precision.[24] Thus, a "community" with some rights to shape the design of this new scientific initiative did emerge, but only to the extent that the community reflected the values of the new science initiative. As a result, although very different in its promises (i.e., that people would have more power to define how their samples would be labeled and used) and practices (e.g., community engagement, ethicist involvement from the very beginning) the HapMap Project, like the Diversity Project, privileged scientific expertise over other forms of authority when conflicts arose. The project thus provided little space for considering how human beings should be ordered and labeled for the purposes of studying their genetic differences—or, put differently, how group differences and group rights should be coproduced. Instead, in many instances the new rights offered to project subjects served only to shore up and legitimate scientists' prior assumptions regarding the naturalness of intergroup divisions.

Rejoining Rights with Responsibility

In both the Diversity Project and HapMap cases, researchers celebrated the effort they made to craft sensitive and inclusive ethical practices along with scientific practices.[25] Yet, as the details of these cases demonstrate, there is no intrinsic value to attempts to forge ethics along with science. Indeed, it may produce unintended problems. Specifically, in both the Diversity Project and HapMap cases, these attempts faltered as they failed to recognize that ethics did not exist separate from the project's "science." Rather, ethical choices inhered in efforts to study human genetic variation, regardless of any explicit effort to *practice ethics*. As a result, in both cases the coproduction of natural and moral orders proceeded in an unreflective manner. Specifically, project organizers failed to account for and respond to the moral and social valuations of human differences implicit in their allegedly "precise" and "scientific" labels, as well as the epistemic choices about how to know human differences implicated in their creation of novel ethical rights for research subjects. A conceptual schema that posited ethics as something that could be added onto science—and not something that was unavoidably implicit in it—impeded organizers' recognition of the value-ladenness of all classifications of human difference. Thus, the construction of new methods of understanding human life became severed from abilities to respond (or response-ability) to how in the process lives would be rendered as different (Haraway 2008, 71, 88).

In the Diversity Project, efforts to produce new rights that might facilitate the grouping of human beings for the purpose of genomic analysis presumed that population-specific human differences already existed, and that Diversity Project researchers could identify and name these groups on scientific grounds. Thus, questions about how groups should form and gain rights in liberal democracies, as well as how human groups can become proper objects of scientific study, remained unasked. By not responding to these bioconstitutional questions about the status of groups as subjects of rights and objects of study, the Diversity Project raised concerns that it would give rise to groups whose status was questionable on both social and natural grounds (Juengst 1998).

Organizers of the HapMap originally planned to remedy these problems in the Diversity Project by providing a process through which the people sampled would have a chance to discuss and decide the grouping and naming of their samples. However, when it became clear that these labels might reflect local and contingent political or cultural values, the project's approach to labeling shifted. In particular, precision eclipsed subject autonomy and engagement as the value that guided the naming process, thus limiting critical reflection on the assumptions that inhere in any act of ordering human differences.

The HapMap's changing policy, and ultimate retraction of the early promise to allow those sampled to decide how to label their samples, left many of the early supporters of the project's community engagement process disgruntled.[26] Some who worked on the project reported that the project's administrators had made a "good faith" effort to efficiently balance rival commitments: on the one hand, to addressing the concerns and views of research subjects, and, on the other hand, to respond to the concerns of HapMap organizers (author's field notes). In the end, however, it was unclear to experienced observers how the resulting HapMap differed from the Diversity Project. In both cases, HapMap organizers set the primary criteria used to define and label groups; these criteria eschewed responsibility for the political and social content of these labels.

Given the overwhelming desire of HapMap organizers to not replicate the problems of the Diversity Project, the question confronting us is how this happened again. Further, what might be done in the future to facilitate greater transparency and self-awareness in processes of constructing human groups for the purposes of genomic research? Comparing the Diversity Project and HapMap cases provides some insights.

First, a simple call for integrating science and ethics is not enough. As theorists of coproduction have attempted to demonstrate, producing a moral order along with a natural order is not a normative mandate (Jasanoff 2004). This is not how things *should* be done, but it is the way things necessarily *are*, independent of actors' specific intent. Thus, the question is not whether science and ethics or rights should be crafted together, but how to respond to and reflectively engage in these bioconstitutional moments. To do this, we must not only understand the dynamics of coproduction, but also follow in detail how coproduction occurs in particular cases—and ask how such processes could be opened up to fuller scrutiny.

Second, scholars and administrators of genomics should pay greater attention to the construction of ethics and rights. For years, scholars of science and technology have developed analytic tools for analyzing how the construction of scientific expertise often eclipses important social and political choices (Hilgartner 2000; Jasanoff 1995; Latour and Woolgar 1986; Wynne 1996). Cases like the Diversity Project and the HapMap highlight the importance of understanding how systems of ethics and rights also play key roles in eliding fundamental social and political issues. Although enrolling ethical expertise might facilitate efficiency, as Jasanoff (chapter 3, this volume) also argues, such an approach does not necessarily engender responsibility for the political and social decisions made in the name of ethical expertise.[27]

Finally, the dominant frameworks in which bioethicists and policy makers ask and answer questions about genomic differences need to expand and become more flexible. Though a focus on questions about eugenics and the possibilities for new forms of biological racism has generated much important work, surveying all research on human difference through the lens of eugenics can have detrimental effects (see also Wellerstein, chapter 2, this volume).[28] One such effect, as the Diversity Project and HapMap cases suggest, is that genome scientists and their administrators seek "precise" methods for ordering human differences and defining groups that they believe protect against bias, namely racism. Such an approach presumes that human groups already exist, and can be precisely represented. Thus it offers no way to consider when and how genetic studies order human beings into groups in a manner that can lead to stigmatization and when and how these acts of ordering illuminate and promote desirable social ends.[29]

Overcoming this limitation and engendering greater collective responsibility for issues of difference requires challenging this naturalized view of difference that inheres in both genomics and liberal democratic conceptions of rights. It is only by recognizing that there are no "precise" correct answers, but that we are always and inevitably caught in the dilemma of determining when and how to recognize and order human differences, that we take the first steps toward reframing rights in a manner that fosters collective responsibility for human genetic differences.

Acknowledgments

Research for this essay was made possible by a grant from the National Science Foundation (NSF #0613026). I thank Sheila Jasanoff and an anonymous reviewer for their astute observations that improved this piece. Any opinions, findings, and conclusions or recommendations expressed in this material are those of the author and do not necessarily reflect the views of the National Science Foundation.

Notes

1. This claim is based on thirty-five semistructured interviews with those who designed and those charged with implementing these new rights. I conducted the first set of these interviews (with organizers of the Diversity Project) from 1996 to 1999. The second set (with organizers of the International Haplotype Map) began in 2004 and continue today.

2. The term "ethical bypass" comes from Donna Haraway's characterization of Sarah Franklin's notion of "built-in ethics" (Franklin 2003). I use the term slightly differently to mean a conceptual structure that leads actors to route their reasoning and decisions around important ethical questions.

3. Feminist legal scholar Martha Minow named this dilemma of determining when "treating people differently emphasize their differences and stigmatize or hinder them" and when "treating people the same become insensitive to their difference and likely to stigmatize or hinder them on *that* basis" the *dilemma of difference* [italics in original] (Minow 1990, 20).

4. A notable exception is the emergence of the debate about the meaning of racial categories in medicine (Schwartz 2001; Cooper, Kaufman, and Ward 2003). These critiques, however, are grounded in claims about population genetics.

5. "Coproduction" is a term created by scholars of science and technology studies to refer to the "constant intertwining of the cognitive, the material, the social, and the normative" (Jasanoff 2004, 6). See, for example, Jasanoff 2004 and Reardon 2005.

6. Organizers of the Diversity Project did think early on about the impact that *The Bell Curve* might have on issues about race and racism their initiative would have to address (Reardon 2005).

7. For a fuller description of the emergence of the Model Ethical Protocol and the group consent provision, see Reardon 2001 and 2005, 98–125.

8. They tie these acts of constructing human difference to corrupt political ideologues, such as Adolf Hitler.

9. For an exploration of the central role the construction of identity and difference play in the constitution of modern forms of power see, for example, Foucault 1976, Melucci 1989, and Butler 1990.

10. The material in this section draws upon previously published work. For an expanded discussion, see Reardon 2001. For biological research, the Tuskegee syphilis experiment is the most notable example. For a history of this experiment, see Jones 1981.

11. The National Commission for the Protection of Human Subjects of Biomedical and Behavioral Research issued the Belmont Report in 1979. The Belmont Report outlines the major conditions of informed consent: consent must be given voluntarily by competent and informed individuals (National Commission for the Protection of Human Subjects of Biomedical and Behavioral Research 1979). *Moore v. Regents of the University of California* (1990) expands the concept of informed to include disclosure of researchers' knowledge of their interests.

12. In this 1978 case, the Supreme Court upheld the right of the Santa Clara Pueblo to enact membership criteria that discriminated against its female members by granting membership rights to children of male members of the tribe who had married a woman not in the tribe, but not to the children of female members of the tribe who had married nonmembers (*Santa Clara Pueblo v. Martinez* 1978).

13. Native attorney in health policy, phone interview with author (January 24, 2001). To address problems created by outsiders defining groups as well as the terms of what counts as adequate protection of those groups, many tribes are now writing their own research codes of conduct. See, for example, <http://www.ipcb .org/publications/policy/index.html> (accessed August 22, 2008).

14. For a statement of this policy, see <http://www.hapmap.org/ethicalconcerns .html.en> (accessed August 22, 2008).

15. Interview conducted by author with HapMap organizer, Washington, D.C. (August 11, 2006); emphasis added.

16. The belief that the HapMap would ultimately allow "communities" to decide the name placed on sample labels circulated widely. Indeed, several otherwise skeptical social scientists cited this commitment to community control as the reason they ultimately decided to join the project. They later expressed disillusionment with the project when it became clear that this policy would not hold (author's field notes).

17. See <http://www.hapmap.org/citinghapmap.html> (accessed February 18, 2008).

18. See <http://www.hapmap.org/citinghapmap.html> (accessed August 22, 2008).

19. Interview conducted by author with project organizer, Washington, D.C. (December 1, 2004).

20. Phone interview conducted by author with project organizer (November 20, 2006). "JPT" is the recommended abbreviation for "Japanese in Tokyo, Japan." See <http://www.hapmap.org/citinghapmap.html> (last accessed February 15, 2008).

21. Interview with author (August 11, 2006).

22. Some would argue that geography has never been a good descriptor of populations, as human beings have always moved around. For examples of this critique, see the physical and biological anthropologists' critiques of geographic descriptors used by some of the Diversity Project organizers (Reardon 2005).

23. Phone interview conducted by author with project organizer (November 20, 2006).

24. For example, one "community" wanted their name written in their native language. This wish was granted (although the label also appears in English). Interview with author (November 20, 2006).

25. HapMap organizers explicitly foregrounded their "integration of science and ethics" in the title of a project publication: "Integrating ethics and science in the International HapMap Project" (The International HapMap Consortium 2004).

26. This dismay in the perceived change in policy was a topic in several of my interviews.

27. For reflections on how the U.S. National Human Genome Research Institute's Ethical, Legal, and Social Issues program explicitly attempted to construct itself around a "scientific" model of peer review and thus establish itself not as a policy body, but as a source of objective expertise in its own right, see Juengst 1996.

28. For a discussion of the problems of understanding genomics within a eugenics frame, see Rose 2007, 58. For a discussion of how "ethics-based critiques of DNA research" have to date crunched "socioethical discussion" of genomics into an overly narrow frame, see O'Malley, Calvert, and Dupré 2007.

29. For an example of the argument that not studying certain human beings will lead to further discrimination and disparities see the arguments of the organizers of the National Center for Human Genome Research at Howard University (<http://www.genomecenter.howard.edu/background.htm> [accessed August 22, 2008]).

References

Biehl, João. 2005. *Vita: Life in a Zone of Social Abandonment*. Berkeley: University of California Press.

Butler, Judith. 1990. *Gender Trouble: Feminism and the Subversion of Identity*. New York: Routledge.

Burchard, E. G., E. Ziv, N. Coyle, S. L. Gomez, H. Tang, A. J. Karter, J. L. Mountain, et al. 2003. The Importance of Race and Ethnic Background in Biomedical Research and clinical Practice. *New England Journal of Medicine* 348:1170–1175.

Cavalli-Sforza, Luca. 1994. "The Human Genome Diversity Project." An address delivered to a special meeting of UNESCO. Paris, France, September 12.

Cavalli-Sforza, Luca, Paolo Menozzi, and Alberto Piazzi. 1994. *The History and Geography of Human Genes*. Princeton: Princeton University Press.

Cavalli-Sforza, Luca, Alan C. Wilson, Charlie Cantor, Robert M. Cook-Deegan, and Mary-Claire King. 1991. Call for a Worldwide Survey of human Genetic Diversity: A Vanishing Opportunity for the Human Genome Project. *Genomics* 11:490–491.

Cohen, Felix. 1982. *Handbook of Federal Indian Law*. Charlottesville, Va.: Michie Bobbs-Merril Law Publishers.

Community Engagement and Sample Collection for the HapMap Project: Background and Overview. 2002. Personal communication with Jean McEwan.

Cooper, Richard S., Jay S. Kaufman, and Ryk Ward. 2003. Race and Genomics. *New England Journal of Medicine* 348:1166–1170.

Dobzhansky, Theodosius. 1962. Comment. *Current Anthropology* 3:279.

Duster, Troy. 2003. Buried Alive: The Concept of Race in Science. In *Genetic Nature/Culture: Anthropology and Science Beyond the Two-Culture Divide*, ed. Deborah Heath, Alan H. Goodman, and M. Susan Lindee, 258–277. Berkeley: University of California Press.

Epstein, Steve. 2003. Sexualizing Governance and Medicalizing Identities: The Emergence of "State-Centered" LGBT Health Politics in the United States. *Sexualities* 6:131–171.

Epstein, Steven. 2004. Bodily Differences and Collective Identities: The Politics of Gender and Race in Biomedical Research in the United States. *Body & Society* 10:183–203.

Epstein, Steven. 2007. *Inclusion: The Politics of Difference in Medical Research*. Chicago: University of Chicago Press.

Franklin, Sarah. 2003. Ethical Biocapital: New Strategies of Cell Culture. In *Remaking Life and Death: Toward an Anthropology of the Biosciences*, ed. S. Franklin and M. Lock, 97–128. Santa Fe, N.M.: School of American Research Press,

Foucault, Michel. 1976. *An Introduction*: Vol. 1. The History of Sexuality. London: Allen Lane.

Hacking, Ian. 1990. Natural Kinds. In *Perspectives on Quine*, ed. Robert B. Barrett and F. Roger Gibson, 129–141. Cambridge, Mass.: Blackwell.

Hacking, Ian. 1999. The Social Construction of What? Cambridge, Mass.: Harvard University Press.

Haraway, Donna. 2008. *When Species Meet*. Minneapolis: University of Minnesota Press.

Hilgartner, Stephen. 2000. *Science on Stage: Expert Advice as Public Drama*. Stanford: Stanford University Press.

Hotz, Robert Lee. 1995. Scientists Say Race Has No Biological Basis. *Los Angeles Times*, p. A1.

The International HapMap Consortium. 2004. Integrating Ethics and Science in the International HapMap Project. *Nature Reviews Genetics* 5:467–475.

Jasanoff, Sheila. 1995. *Science at the Bar: Law, Science and Technology in America*. Cambridge, Mass.: Harvard University Press.

Jasanoff, Sheila. 2004. The Idiom of Co-Production. In *States of Knowledge: The Co-Production of Science and Social Order*, ed. S. Jasanoff, 1–12. London: Routledge.

Jensen, Arthur. 1969. How Much Can We Boost IQ and Scholastic Achievement? *Harvard Educational Review* 39:1–123.

Jones, James H. 1981. *Bad Blood: The Tuskegee Syphillis Experiment*. New York: The Free Press.

Juengst, Eric. 1998. Group Identity and Human Diversity: Keeping Biology Straight from Culture. *American Journal of Human Genetics* 63: 673–677.

King, Mary-Claire. 1996. Relevance of the Human Genome Diversity Project to Biomedical Research. Retrieved from <http://www.stanford.edu/group/morrinst/hgdp/MCK2NRC.html> (accessed June 11, 2001).

Latour, Bruno, and Steve Woolgar. 1986. *Laboratory Life*. Princeton: Princeton University Press.

Lewin, Roger. 1993. Genes from a Disappearing World. *New Scientist* 1875:25–29.

Livingstone, Frank. 1962. On the Non-Existence of Human Races. *Current Anthropology* 3:279.

Marks, Jonathan. 1998. Letter: The Trouble with the HGDP. *Molecular Medicine Today* 4 (6): 243.

Melucci, Alberto. 1989. *Nomads of the Present: Social Movements and Individual Needs in Contemporary Society*. Philadelphia: Temple University Press.

Minow, Martha. 1990. *Making All The Difference: Inclusion, Exclusion, and American Law*. Ithaca, London: Cornell University Press.

Moore v. Regents of the University of California. 1990. 51 Cal.3d 120, 793 P.2d 479, 271 Cal.Rptr. 146.

National Commission for the Protection of Human Subjects of Biomedical and Behavioral Research. 1979. *The Belmont Report: Ethical Principles and Guidelines for the Protection of Human Subjects of Research*. Washington, D.C.: Office for Protection from Research Risks, National Institutes of Health, Department of Health, Education and Welfare (April 18).

North American Regional Committee, Human Genome Diversity Project. 1997. Proposed Model Ethical Protocol for Collecting DNA Samples. *Houston Law Review* 33:1431–1473.

O'Malley, Maureen, Jane Calvert, and John Dupré. 2007. The Socioethical Study of Systems Biology. *American Journal of Bioethics* 7 (4): 67–78.

Proctor, Robert. 1988. *Racial Hygiene: Medicine under the Nazis.* Cambridge, Mass.: Harvard University Press.

Rabinow, Paul. 1999a. Artificiality and Enlightenment: From Sociobiology to Biosociality. In *The Science Studies Reader*, ed. M. Biagioli, 405–416. New York: Routledge.

Rabinow, Paul. 1999b. *French DNA: Trouble in Purgatory.* Chicago: University of Chicago Press.

Reardon, Jenny. 2001. The Human Genome Diversity Project: A Case Study in Coproduction. *Social Studies of Science* 31:357–388.

Reardon, Jenny. 2005. *Race to the Finish: Identity and Governance in an Age of Genomics.* Princeton: Princeton University Press.

Rose, Nikolas. 2007. *The Politics of Life Itself: Biomedicine, Power and Subjectivity in the Twenty-First Century.* Princeton, N.J.: Princeton University Press.

Santa Clara Pueblo v. Martinez. 1978. 436 U.S. 49.

Schiebinger, Londa. 1993. *Nature's Body: Gender in the Making of Modern Science.* Boston, Mass.: Beacon Press.

Schwartz, R. S. 2001. Racial Profiling in Medical Research. *New England Journal of Medicine* 344:1392.

Sharp, Richard, and Morris Foster. 2000. Involving study populations in the review of genetic research. *Journal of Law, Medicine & Ethics* 28:41–51.

Steinberg, Douglas. 1998. NIH Jumps into Genetic Variation Research. *The Scientist* 12:1.

Venter, Craig. 2000. Statement on Decoding of Genome. *New York Times*, June 27, D8.

Wilkins, David. 1997. *American Indian Sovereignty and the U.S. Supreme Court.* Austin: University of Texas Press.

11

Despotism and Democracy in the United Kingdom: Experiments in Reframing Citizenship

Robert Doubleday and Brian Wynne

At the turn of the millennium, a series of fiascos over scientific advice to government challenged the peculiarly British ways in which such advice had been procured, framed, and used. Prominent episodes included controversy over decommissioning the Brent Spar offshore oil facility, resulting in Greenpeace's victory over the UK government and Shell in 1994; and the crisis over the UK government's handling of BSE (bovine spongiform encephalopathy, or mad cow disease) which came to a head in 1996 (Grove-White 1997; van Zwanenberg and Millstone 2005). Such challenges to the presumptive authority of scientific advice over public policy and public life proved fertile ground for controversy over genetically modified (GM) crops over several years straddling the millennium. These crises expressed and intensified what we argue has amounted to a constitutional unsettlement of relations between the state, science, and citizens in Britain.

Developments in the twentieth-century roles of science in government—developments that, in keeping with the theme of this volume, we would call (bio)constitutional—led in the United Kingdom (and elsewhere) to a turn-of-millennium condition in which science had become deeply entrenched as *scientism*. Scientific advice and authority were being systematically exaggerated in regulatory control and public debate, as in the regular use of risk assessment for public reassurance, as if that reassurance were based on science's full independence from policy commitments and assumptions. Further, as scientific advice took on a greater role in post Second World War public policy, it became by default not only an *informant* of public policy (its classical role), but also a powerful *cultural* agent, as arbiter of *public meanings*. This extension of science into *scientism* was not a consequence of deliberate design but rather of mutual accommodation and mutual reinforcement between policy and science as institutionalized epistemic (and hermeneutic) authority. Thus science assumed the role of authoritatively providing the *meaning* of many public issues, which came to be defined as "risk issues" or even "scientific issues," obscuring other key dimensions.

The shocks of the Brent Spar, BSE, and GM controversies provoked a significant shift in scientific governance in Britain, marked by an explicit concern for building public trust through greater openness to public scrutiny and participation. Foremost among the many articulations of the British state's newfound concern for public engagement in science were a series of interventions by Parliament, the executive, and the judiciary in the year 2000: the House of Lords Science and Society report; the establishment of the Food Standards Agency, as well as inclusive standing commissions covering human genetics and agricultural biotechnology; and the Phillips Inquiry into the BSE crisis (Phillips 2000). This shift condensed around a set of practices for eliciting public views through formal techniques of "public engagement." A departure from past governance approaches, this potential space for democratizing science policy was itself shaped by British political institutions and civic epistemologies—negotiating consensus around empirical demonstrations of public attitudes (Jasanoff 2005, 247–271).

Alan Irwin characterizes these developments as actualizing a new governance of science in which public trust is equated with social consensus, achieved through official modes of public participation. Irwin (2006, 303) rightly calls for greater effort to be paid to describing these shifts in governance as forms of social experiment "symptomatic of the contemporary culture of scientific and technological change." In this chapter, we contribute to this goal by focusing on how citizenship was reframed through this partial realignment of governance with respect to the science, technology, and innovation politics of genetic modification.

Our central argument is that the recent history of British policy toward public engagement with science can be described as a playing out of tensions between competing versions of the place of citizens in shaping public meanings (and thus also, material trajectories) of what is at stake in a "knowledge economy." We characterize these contending versions of the capacities of citizens as "despotic" or "democratic" with respect to citizens' rights to participate in the production of "public objects." By this, we mean the extent to which technoscientific policy choices such as R&D and innovation trajectories embody tacit values established by democratic collective action (see also Jasanoff, chapter 1, this volume, and for a liberal democratic history of the concept, see Ezrahi 1990).

The chapter considers three instances in which British institutions confronted and creatively responded to citizen articulations of alternative understandings of GM and the issues at stake; that is, citizens expressed different meanings and concerns from those recognized by the British state and its supporting science. Following Jasanoff's seminal account of British communitarian civic epistemology, we note that tensions between despotic and democratic constitutional commitments retain long-running "British" political cultural attributes of empiricism, instrumentalism, informality, and pragmatism, aversion to models and abstract thinking ("speculation"), and of consensus by enrolling supposedly innocent publics in projects of "common vision" (Jasanoff 2005).

Notwithstanding its long-established parliamentary politics, the UK state bears continuing witness to its deeply monarchist traditions, in which, for example, the rights and ingrained sense of agency of *citizens* are circumscribed by their self-conscious standing as (the monarch's, and thereby de facto the state's) *subjects*. Although healthy traditions of truculent independence are never deeply buried, the postwar rise of science as public authority is aligned with the centralizing tendencies of the British state and a relatively passive position for its subjects. Classic manifestations of this alignment can be found not only in the case of nuclear power (Wynne 2011), but also in other state-sponsored high technologies, such as biotechnology in the 1980s and 1990s. The constitutional implications of this settlement, we argue, are in the strict sense despotic, in that there is no place for public debate about the meaning of the policy issues at stake. By "despotic," we mean that state and scientific modes of practice in policy domains pervaded with scientific-technical dimensions increasingly assert that the *meaning* of the public issue is adequately framed by technoscience, and there are no publics who might bring different, relevant meanings and concerns to bear. In effect, citizens play a role on condition of alignment of their meanings with those already laid down by science and the state. This effectively means that there is no public to be considered in defining modes of policy action or commitment—hence, de facto, no polis.

We describe how the genomics revolution and agricultural biotechnology in particular entered a field of contradictory forces over how to arrive at shared meanings of technoscientific innovation. On the one hand, closure is reached through the state's deference to science and technology as default authors of public meanings of innovation (both its risks and its promises); on the other hand, citizen unrest, including dissenting scientific voices, have opened up space for contesting those hegemonic tendencies. These moves have tentatively recognized civic rights in relation to science, in which citizens are recognized as legitimate authors of the public meanings which science, as a vital component of modern democracies and agent of the state, should respectfully negotiate with, as well as inform (Arendt 2005).

In this chapter, we suggest that the developments in 2000 outlined previously mark epochal moves to open up to civic accountability the largely unaccountable late-twentieth-century convergence between the British state, science, and global capital. This shift is exemplified by official commitments to models of public dialog to inform the growing range of public policy issues involving technoscience, but unevenly retaining the deeper structuring impositions of public meaning as explained previously. In these, as Wynne (2006) has observed, public differences of *meaning* were understood only as inadequacies in knowledge. Our argument here is that the politics of contemporary British science advice, regulatory policy, and public engagement with science can be understood as the expression of deep contradictions between more despotic and more democratic models of state-science-citizen relations. Nowhere have these dynamics been more apparent than in the debates over agricultural biotechnology—more specifically, GM crops and foods.

Rewriting Citizens through the Nonhuman Genetic Sciences

We take in this section three cases in which public questions were posed concerning the fundamental biological and normative processes through which the "texts of Life" are being rewritten into GM agricultural plant organisms and their definition, assessment, and promotion. In these scientific, policy, and public processes, we see the coproduction of biological activities with embedded understandings of civic capacities and rights, especially in relation to scientific knowledge itself. As we will show, these understandings were deeply normative. Indeed, one major shift to which the genomics revolution has been central is from scientific research seen as understanding biological processes, putatively prior to imagined commercial technologies, to scientific research that already embodies explicitly those imagined sociotechnical ends, as if these were also natural.

This altered balance in the epistemic culture of technoscience involves itself a normative shift in the constitutional structure of agency and responsibility between science and democracy (the representation of citizens). This is because the conventional definition of the responsibility of a democratic society is that it—and not science—should choose which imagined applications are acceptable and desirable from whatever range of possibilities that science *discovers* in nature. A largely concealed form of normative intervention by science into democratic society is thus seen to be in play here; the typically more informal, empiricist, and incrementalist constitutional fabric of the United Kingdom may be more significantly vulnerable to enabling such shifts than others with more wide-ranging traditions of constitutional debate.

This constitutional role of science in writing life through various fields of genetics and related disciplines includes the attempted articulation of ontological categories that would shape the ethical, risk, and political reactions of the UK public at large. When organisms whose genetic code has been engineered in order to instill commercially desired crop traits are declared by government and its regulatory science advisers to be otherwise identical to the original organism, then the consequent ontological *diktat* logically allows for no conceivable unpredicted consequences from the release of such organisms into the environment. Yet European citizens' concerns over GM crops and foods were motivated by questions such as which authority would recognize and be responsible for unpredicted consequences (Marris et al. 2001; Wynne 2001). These very general public concerns were emphatically though tacitly disallowed and were translated instead into fundamentally different representations as (exaggerated) anxieties about (by definition) known risks. Not surprisingly, therefore, these excluded civic reactions were amplified by concomitant reactions against institutions acting with such normative commitments in the name of objective science.

Against the background of moves toward opening of citizen relations to science and the state outlined earlier, we use the case history of public engagement in the UK GM issue to show how UK citizens and "the public" itself were cast as political subjects through performances by the scientific and policy agencies of the state. This involved attempts to define presumptively—through the "natural" authority of science as privileged by the state—not only what knowledges and knowledge-(in)capacities, but also what meanings, values, and ethical identities such public subjects supposedly did, and should properly, have.

As a period in public engagement with science that stretched, episodically, from about 1992 to 2004, the GM issue in the United Kingdom was both institutionally organized and spontaneous at different stages of an uneven public life of controversy and confusion. This same period also witnessed, and the events themselves were the dynamic fabric of, a basic shift in the informal constitutional relationships between the UK state and its citizens, mediated by science. In particular, public opposition to prevailing state definitions of the issues challenged what had become the state's inadvertently embedded practical presumption, that science could not only inform public policy commitments, but could give such commitments their effective public meaning, as if they concerned risk (hence scientific) issues only. As we show later, more loosely articulating state bodies, such as research councils, showed signs of recognizing this self-inflicted institutional misunderstanding. However, more centrally placed government departments and agencies showed few such signs, even while their autocratic attempts to define public meanings, benefits, and values were manifestly failing.

Here we review developments in what has been called the "public understanding of science" and more recently "public engagement with science" since the landmark report of the UK House of Lords Select Committee on Science and Technology in March 2000. The Lords' report was a specialist product, in that a majority of the Select Committee were eminent scientists and engineers, all of whom had played influential roles as senior government insiders and advisers, and some of whom had been on the receiving end of controversy and critique. For the first time, an official institutional voice expressly repudiated the institutional habit of the UK government to presume that public differences with authority over matters involving science and policy must be due only to public ignorance of the science—the so-called public deficit model (Wynne 1991, 1992, 1993), which had been criticized as having no evidence to support it and as erroneously identifying the evident and undenied existence of public ignorance of much of science as the *causes* of public difference and dissent. Rejecting this entrenched deficit model as patronizing and self-defeating, as well as lacking in evidence, the Lords' report called instead for two-way public engagement with science to become a routine governmental practice (House of Lords Select Committee on Science and Technology 2000). Thus replacing root-and-branch the paternalistic one-way, corrective public education that previously typified UK policy and scientific culture (Royal Society 1985).

Over the intervening ten years since its publication, this authoritative parliamentary body's move has at least ostensibly[1] become more or less institutionalized, if variably so, within the state's administrative structures and practices, and also in more adventurous form perhaps, in the quasi-independent UK research councils. As already indicated, there were independent developments that pushed in the same dialogical, opening direction, but in condensing many of those separate forces and feelings into a powerful institutional statement, the Lords' report has become something of an iconic reference point—a status that perhaps exaggerates its actual impact.

Unilever: Recent UK History of Public Engagement with Science

Evolving UK policy processes showed contradictory tendencies toward both closing down of public meanings of GM innovations, through an exaggerated state deference to science, and also opening up in response to self-assertive counterreactions at grassroots political levels by widely and variably based citizen networks. In this respect, Britain enjoyed vibrant, well-supported, well-informed, and effective civil society groups and social movements intervening in creative ways in policy processes, including in their scientific domains. Moreover, some agencies of the state and institutions with state-like attributes that were less centrally connected with the government machinery were more receptive to civil society networks. Contingent connections meant that some institutions were open to public dialog and public engagement, reflecting a significant and indeed (for the UK state) unusual degree of recognition of those spontaneous public counterforces at the heart of science policy itself. Thus, even when the closing down processes appear to reflect a coherent and unitary entity in the state, they resemble more a portfolio of less than consistent forces and interests whose overall orientation remains to some degree open to negotiation.

Early signs of the potential constitutional significance of the GM controversy were arguably picked up by industry before they were by the state. In 1994, Unilever began to take seriously the potential public opposition to dominant meanings attributed to GM by scientists and the state. Unilever, a major global food company, at that time had interests in agricultural biotechnology research and development. Seeking to apply lessons from a formalized stakeholder dialog in the Netherlands around the use of GM bacteria in the production of "biological washing powder," key company officials initiated an informal engagement process with environmental and consumer nongovernmental organizations (NGOs) in Britain (Doubleday 2004).

In 1994, Unilever and the environmental NGO Green Alliance co-hosted a seminar in London to which NGOs representing a wide range of consumer and environmental interests were invited. Following this event, Unilever and Green Alliance established a small but open "Contact Group" between representatives from the firm and NGOs. The Contact Group held ad hoc meetings at intervals of about six months over the next seven years to discuss issues concerning the development, regulation, and commercialization of GM technologies. The Contact Group operated informally, and close working relations developed between Unilever and the British NGO community over the issue of GM foods, lasting beyond the group's final meeting in the summer of 2001.

Participants in the Contact Group have stated that one of the most valuable consequences of their discussions was the commissioning of social science research.[2] Two of these studies explored public attitudes to biotechnology. One of the Contact Group's principal members hinted in passing at the exploratory and experimental spirit of these novel interactions: "We were very fortunate to find a common interest with the Green Alliance and members of the NGO contact group: to understand better what was going on, particularly from the perspective of the public as consumers and citizens, and to invest time and effort to put up the questions and hypotheses for deeper study" (Schofield 2002).

Within Unilever, the Contact Group became an important mechanism for incorporating a wider range of perspectives in the company's reflections on the development and commercialization of GM foods. The interaction with social scientists helped some people within Unilever to reframe how the company conceptualized public attitudes to innovation. This reframing was also used within the company to reflect on public attitudes to the social and environmental implications of corporate practice. Within such a large and global company, the influence of this sustained interaction with NGOs and (avowedly interpretive) social science on corporate attitudes remained uneven.

Unilever's NGO dialog through the Contact Group was used by the company to respond to a controversy that could not be adequately managed within the corporation's established models of market research, product development, government affairs, and public relations. Unilever participants in the NGO dialog began to use the terms "consumer-as-citizen" or "consumer-citizen" to describe public attitudes to GM foods that could not be captured in terms of consumer preferences alone. According to this understanding, Unilever believes it needs to take account of citizenlike expressions and meanings on the part of consumers—toward the technologies used in its products, and also toward the corporation itself.

The politicization of the act of consumption captured by the term "consumer-citizen" is interestingly different from the concept of "citizen-consumer," which expresses the marketization of civic and political relations (Iles 2004). The term "consumer-citizen," by contrast, takes the corporation as its analytical starting point. Unilever's NGO dialog provides an opportunity to study the emergence of the category of consumer-citizen as it opens up a new domain in which citizenship rights are negotiated in relation to a private body rather than the state.

The practices of negotiation of expertise and representation of publics that led Unilever to reframe its relations to its consumer-citizens led to a potential opening up of the firm to wider democratic input and accountability. In order to frame public attitudes in these terms, Unilever needed to institute techniques to aggregate and represent public attitudes; in doing so, the company became embedded in networks of NGO and academic practice. Tensions became apparent in the deployment of the term "consumer-citizen," indicating that the company's institutionalization of practices to take account of consumers' citizenlike expressions were necessarily incomplete and open to challenge from outside the company. Yet these moves indicated that the company was recognizing an institutional responsibility to understand and respond to citizen concerns that went well beyond conventional market definitions of corporate responsibility and that in practice, existing state institutions were not themselves acknowledging (Grove-White et al., 1997, 2000; Marris et al. 2001; Wynne 2001, 2006).

As Unilever tried to cope with public controversy over GM technologies in the UK, it extended the meaning of citizenship beyond its traditional political definition. Unilever's engagement with NGOs and its subsequent commissioning of social science research on public attitudes to GM foods can be understood as an attempt to reframe the company's relationship with its publics. In doing so, it followed tried and tested patterns of identifying relevant experts and negotiating consensus through small-scale empirical demonstrations and direct informal interactions—reflecting long-running habits of British civic epistemology. In responding to a crisis of legitimacy through seeking to represent and respond to publics, Unilever ended up acting as a quasi-democratic quasi state. In key respects, this reflected another tradition in UK policy culture, in which big private corporations have enjoyed close identification with state interests and needs, and have thus absorbed some of that UK sense of paternalism, as well as an internalized sense of responsibility for state functions such as maintaining public legitimacy and, ultimately, social order.

As it happens the careful work of collaboration, research and reflection on the part of a small number of key Unilever personnel was overtaken by events when Monsanto commercialized its GM soya crop in 1996—against strong representation by Unilever's chairman. However, the recognition that public attitudes could not be reduced to questions of risk alone were lodged in what was a quasi-state institution, prepared to grant greater seriousness to arguments made by individuals such as Sue Mayer, Andy Stirling, Robin Grove-White, and Brian Wynne, all of whom had been involved in Unilever-funded research through the Contact Group and who went on to play a role in the elaboration of the democratic model of engagement with GM in the UK (Grove-White 2001).

GM Nation?: Democratic Debates and Despotic Responses

Unilever's NGO dialog took place in a context where new exercises in public engagement with science were proliferating, with parliamentary and ministerial endorsement. The Royal Society's establishment of a Science in Society advisory committee in 2000, including social scientists and others leading these developments, gave institutional authority to these moves. Punctuating this decade, however, was the biggest episode in two-way public engagement with science yet to have been deliberately and officially conducted in the UK—indeed, maybe anywhere in Europe. This was the troubled and intense *GM Nation?* public debate orchestrated by the short-lived Agriculture and Environment Biotechnology Commission (AEBC). The multidimensional deliberative exercise succeeded several years of trenchant and unresolved—indeed, escalating—public controversy: an earlier (1994) consensus conference conducted by academics for the Biotechnology and Bioscience Research Council (BBSRC); various forms of direct action against GM crops; jury acquittal of the protesters from criminal charges; criticism of official scientific claims; colorful media campaigns; successful, high-profile public critiques of scientific inputs to state policy; and food retail sector initiatives to reflect public attitudes in sourcing and selling non-GM foods. All of these expressions of dissent from the clear and convinced preferences of the state undermined the public legitimacy of the institutional fabric supporting the government's pro-GM standpoint.

Only a year after its establishment in 2000, the AEBC had itself challenged government ministry statements that science—in the form of expected findings from the unprecedented 2000–2004 farm-scale evaluations of GM crop production for biodiversity effects—would provide the definitive answer to the political question "GM crops or not in UK agriculture?" Yet, in the face of escalating controversy, the government could still see only one way to trump political difference—more science. The issue was defined repeatedly as being only about scientifically based risk, a position emphatically endorsed in a major speech by Prime Minister Tony Blair in 2002 at the Royal Society (discussed in Irwin 2006, 308–309). Thus, while on a more general front opening moves toward society were being explored, as encouraged by the House of Lords' 2000 report, in the case of GM, state institutions drove in the opposite direction toward greater insistence on their own uses of science, not only to inform policy and the public, but also as author of public meanings.

Embarrassed by the AEBC's public assertion that this scientism was not good enough (AEBC 2001) and reeling from other effective attacks by critics of GM policies, the UK government agreed to the three-pronged public deliberation exercise *GM Nation?* that was composed of a full blue-ribbon GM Science Review, chaired by the government chief scientific adviser; a study of the economic benefits of GM crops for the UK, conducted by the Prime Minister's Strategy Unit; and a nationwide public engagement designed, conducted, and reported upon by the AEBC. These three prongs were originally supposed to be coordinated and cross-fertilizing, but in practice this proved too demanding.

AEBC's establishment in 2000, along with two other bioknowledge- and biopolicy-focused "supercommissions," on food (the Food Standards Agency) and on human genetics (the Human Genetics Commission) had been an explicit government response to public and NGO disaffection, reverberating in the media, with felt industry capture and undue narrowness of the investigatory agenda of conventional scientific advisory committees (de facto, regulatory decision-making committees). Combining, as it did, ardently pro-GM scientific and industrial advocates with demanding scientific and NGO critics, the AEBC had in its short existence come to be a microcosm of the larger conflict—except that it was more difficult within the confines of the AEBC to dismiss the science offered by critics as simply false science, in the way that the state could and did in the wider public domain. The AEBC was terminated by the government in 2004, possibly because its handling of GM-related policy issues was seen as closer to wider public concerns than mainstream pro-GM actors were willing to allow.

The *GM Nation?* public engagement was designed and evaluated in part by academic social scientists (Horlick-Jones et al. 2007. It involved extensive preliminary social research by a consultant group, Corr-Wilburn, as well as many (more than six hundred) public meetings of variable format, and spontaneous self-motivated participation. These meetings also included a simple self-response attitude survey, centrally collected and analyzed; further narrow but deep structured explorations of public attitudes; and a final analysis and report. The report's conclusion was that, notwithstanding recognition of potential benefits, public attitudes remained in the main resolutely against commercial GM crops and foods, at least for the United Kingdom. This unwelcome result for the government, which remained determined to attract global GM industry research investment to Britain, provoked a backlash that effectively alleged incompetence and even anti-GM bias on the part of the AEBC. The claim was that the public meetings had not allowed a real silent public majority to speak but had instead been infiltrated and overwhelmed by self-selecting minority opponents of GM (Lezaun and Soneryd 2007). Social science analysis tended to support this critical evaluation, partly because of the mainstream methodological selection bias in favor of "neutral" participants for such exercises, which tended to disqualify anyone with prior interest in the issue as if they were not proper citizens.

The government's long-awaited April 2004 decision about licensing commercial GM crops in Britain was something of a compromise position. It made no reference to the AEBC report of public refusal, but in accepting the extant Bayer Crop Science Bt GM licensing proposal, it laid down restrictive conditions, advocated by the AEBC, that Bayer said made the commercial planting unviable anyway. Thus commercial GM crops remained unplanted in the United Kingdom, even though officially accepted by the government.

Later EU-initiated commercial licenses for GM crops have also not been enacted in the United Kingdom, despite their legality in principle. Moreover, in 2008, GM potato crop trials at Leeds University were ruined by protesters, indicating a continuing readiness on the part of some public groups to take direct action to further what public surveys show to be majority public opposition to such crops in the United Kingdom.

However, pervading and coloring these partially opening moves and—individual exceptions notwithstanding—persisting when it was tested in the important public issue of GM crops and foods, the deep institutional cultural reflex was that the general polis was incapable of proper knowledge about issues that the state and its agents repeatedly defined as solely "scientific" issues like risk (and moreover, "risk" only as restrictively defined by state scientific bodies), and that this public incapacity was the reason for their difference from government and its science-authority (Wynne 2006). The typical UK public continued to be defined as incapable of "making sense" properly—an official stance in which factual and normative questions were inextricably intertwined, in contradiction of the state's own assertions that facts were separable from values, and that only facts—those of the state, that is—were salient. One senior government scientific adviser even complained that the public needed to be educated to understand properly the benefits of GM crops, which he took for granted as a scientific fact (Burke 2004). It is significant that polls showed that the public did recognize that GM crops had *potential* benefit—the crucial issue that was ignored was the further public skepticism as to whether, and if so, under what conditions, those possible benefits might be realized in practice, under the prevailing conditions of corporate concentration in scientific research, innovation, and control of seed distribution in global agriculture and trade.

GM Nation? was never allowed to reach the point of understanding and representing typical public concerns about GM. Nor, despite its efforts, did it succeed in escaping the longer-standing process of imposing on the UK public a state-imagined set of (in)capacities that then correspondingly delimited civic rights in relation to scientific authority and the political commitments lying behind it. As already intimated, science is an extremely important mediating—and multidimensional, often ambiguous—actor in these relations. In effect, in such issues, science has been handed the sovereign task of defining public meanings, not only of informing them. Therefore, state-citizen relations are necessarily affected by the roles and meanings given to science by state bodies, including ones that are themselves defined as scientific, like regulatory advisory committees. Throughout the UK GM controversy, the intrinsic contingency of scientific risk knowledge was denied by the state, and the predictive power of science exaggerated. This denial of contingency, as well as the unstated extension of scientific agency from informative knowledge into the different realm of public meanings, imposed corresponding limits on the recognized capacity and agency of citizens. This despotic shift—unnoticed, unstated, probably inadvertent, and thus nonnegotiated—was not a sharp and dramatic political ploy, but it was no less arbitrary. Such processes of denial, and the actors they implicate, are endemically open and incomplete. Nevertheless, there are distinct and maybe distinctly British patterns that we can attempt to identify and analyze beneath the surface ferment of normal political activity and commentary.

There is one particular syndrome that we have highlighted in these evolving processes in Britain. The House of Lords Select Committee's key 2000 proposal was a genuine move to redefine and enlarge civic rights with respect to state-organized and state-orchestrated science, especially science for public policy. This represented a definitive move to make government science not only preach to its publics, as in the hitherto entrenched presumption of the need for one-way, corrective, education of an ignorant and *normatively mistaken* public on matters involving science, but instead to listen to citizens, in two-way dialog. Yet this move, ostensibly genuine in principle, harbored deeper institutional and constitutional complexities that were not at the time fully articulated nor understood.[3] They thus left wholly undefined what such fundamentally novel requirements of listening to citizens might mean in practice.

Adventures in Public Engagement and UK Plant Sciences?

Amplifying the UK institutional inhibitions against a more active role for civic actors in public issues involving science have also been fears as to what public engagement might involve. The academic advocacy of "upstream" public engagement, as distinct from engagement only on downstream issues of risk and impacts, was accompanied by the crucial, explicit note that this shift of attention meant asking different questions, in a different way, about the meanings, purposes, and priorities imagined for scientific research. When this upstream shift was endorsed by the enthusiastic GM-promoting Minister for Science Lord Sainsbury (Sainsbury 2004), and indeed by the Royal Society's 2004 *Nanoscience* report (Royal Society and Royal Academy of Engineering 2004), that original understanding was transformed and retrenched back into the idea that upstream was simply about earlier anticipation of risks and impacts and earlier warning of public concerns, at a point where they might be preempted before they became disruptive. The original intent was thus overlooked and unnecessary fears were fueled that upstream dialog meant allowing the public to dictate to scientists what research they should be doing next, on what, and in what ways.

Almost unnoticed, a much more authentic, scientifically informed, and reflectively reasoned example of upstream public engagement with science—through indirect, scientifically mediated *reorientation* of the imaginaries shaping scientific research and its funding—has occurred in the UK GM controversy and related science policy. This is indeed modest and fragile, but also potentially groundbreaking. Here we need to shift focus to the UK BBSRC as the main public funder of basic and strategic plant and crop science.

As has been recognized (van Lente 1993; Jasanoff and Kim 2009), forward-looking imaginaries embodied and enacted in public funding and research for science and technology are a key factor shaping the directions and outcomes of innovation. Significantly, these shaping imaginaries are invariably social as well as technical, in mutually reinforcing forms, and they are typically underarticulated in public reasoning, even if they are influencing the forms and directions of major social consequences and inadvertently excluding other unimagined or neglected possibilities.

From roughly 1990 to the late 2000s, an inadvertent selective commitment had become cumulatively entrenched in the social-institutional structures shaping UK agriculture and in the UK BBSRC (previously the Agriculture and Food Research Council). This imaginary had materialized and developed in a somewhat monolithic and concentrated form, with attendant loss of intellectual and stakeholder diversity. The concentrating forces were not exclusively GM-oriented interests and expectations but also included commitments to particular model plant species and exclusive dependencies on big technologies; further factors were GM-related, but indirect, as explained shortly. Nevertheless, the pervasive expectation that GM crop and food futures would dominate global innovation and markets had pervaded the plant sciences as elsewhere, and many public scientists were sincerely convinced spokespersons for this imaginary. Until the public opposition to GM crops condensed resoundingly into *GM Nation?*, this was the uniformly entrenched but implicit commitment driving BBSRC's research and related policies. Yet in 2004, the BBSRC, as an institution of the state, recognized the legitimacy and salience of public opposition to GM organisms and of the uninvited public engagement that had been expressed through the persistent and proliferating controversy. Such engagement, upstream in key senses, but also long downstream in terms of the formulation of government priorities, reshaped the monolithic, institutionally entrenched, sociotechnical imaginary, in material (if still fragile and open-ended) ways.

During the late 1990s and early 2000s, the BBSRC was publicly committed to promoting scientific research on the assumption that UK agriculture would be fully integrated with a global vision of agriculture in which GM crops would be grown commercially and traded globally. The BBSRC's chief executive, Ray Baker, himself an ex–chemical industry executive, appeared frequently on radio and television arguing the pro-GM case against NGO critics and academic skeptics. Such imaginaries of monolithic concentration, in which GM was simply assumed as *the* future for agriculture, seamlessly integrated GM-oriented, publicly funded science with international corporate R&D and its structures of intellectual property rights.

Thus, when—between 2002 and 2005—social and natural sciences researchers (Stengel et al. 2009; Kamara 2009) interviewed many UK university and BBSRC scientists about the effects of intensified commercial demands on their scientific research, a repeated lament was about the undue concentration and narrowing that had occurred over an extended period in UK plant sciences, including plant genomics research. Not all of this was due to a singularly GM vision, but GM was a central element. In one 2003 example, an internationally leading crop plant geneticist interviewed at the BBSRC's John Innes Research Centre complained that other non-GM approaches (such as using plant genomics to assist conventional plant breeding, as with marker-assisted selection and breeding) were being ignored and starved of public funds to the benefit of GM-oriented science, which he described as a scientifically inadequate commercial "short circuit from lab to field."

A further tangential agricultural policy development a few years earlier had already reinforced the GM trajectory concentration in science itself. In the late 1980s, the Thatcher government had privatized the national Plant Breeding Institute at Cambridge, in the process undermining public plant breeding as a diverse community of commercial and knowledge production (Webster 1989). Thus, an important institutional vehicle that could feed farmer interests and knowledges back upstream into research, similar in some respects to the U.S. agricultural extension program, was destroyed. Without this field-grounded stratum of expertise and possible mediation, laboratory plant sciences were left with the only route from lab science to field as the high-science, high-technology GM "short circuit," from commercially oriented R&D lab to the industrialized, commercially regulated field. Agricultural research was thus constrained in its imaginable crops and in the organizational modes of such narrowly selected crop production (van Dooren 2009). This is a classic empirical manifestation of the STS concept of "coproduction" (Jasanoff 2004). Science and society were mutually shaped not as a direct function of specific decisions, but through a longer more gradual process of "naturalization" of commitments and adjustments between interdependent and interactive agents.

It is interesting to note that the BBSRC stimulated a process of self-reflexive learning as to its own strategic scientific commitments. In 2003, the BBSRC initiated a crop science review. This scientific community review shifted plant and crop science policy in important (if modest) ways that have yet to be more widely recognized. The resulting *Crop Science Review* (BBSRC 2004), which included a consultation with relevant scientists and stakeholders-users, stated a marked if imprecise shift in scientific ethos: "[the review] proposes a stronger national focus on research underpinning 'public good' plant breeding" (6).

The BBSRC defined public-good plant breeding in terms that explicitly noted the tie-in between good science and good politics: "***Public-good plant breeding***: The consultation exercise identified a widely perceived need for public-good plant breeding, in order to address crops and traits not emphasized by multinational interests and to restore public confidence in plant breeding" (6; emphasis in original) The review concluded more specifically and directly: "BBSRC should seek to increase publicity for public-good plant breeding and to emphasise the role of genomically-informed but non-transgenic approaches to crop science research" (11). And further: "We highlight . . . the need to identify and generate new sources of variation for important traits and to strengthen the science underpinning the development of non-GM approaches to crop improvement. . . . Transgenic plants involving gene transfer between species are not a prerequisite for exploiting genomics" (23).

This was not an anti-GM move. It recognized the routine scientific view that transgenic techniques are a useful (but not exclusive) tool for understanding gene function. However, the review did display an important shift in the BBSRC's sociotechnical imaginary to a more holistic, diversely grounded, and flexible portfolio of future scientific, agricultural and social possibilities and priorities. It is worth noting that although this upstream shift did occur as a result of public intervention[4]—largely indirectly—in GM science, reflection by scientific actors on the reasons for those public interventions, and their recognition that technical alternatives could be adopted in response, were also crucial. No typical nonspecialist member of the public would have been able to conceptualize and propose these strategies for themselves. Since the BBSRC *Crop Science Review* was published and digested, modest material moves have been made according to this new science policy imaginary, with £15 million being spent by BBSRC in 2008–2009 in the ensuing Crop Science Initiative. Where this may lead remains open.[5]

Conclusions

In this chapter, we have described several key developments in UK public engagements with science that bear upon and reflect the changing relations of British state, science, and citizenry. Interestingly enough, in terms of a comparative analysis of national cultures of risk, science, and governance, big corporate actors like Unilever expressed many of the same key distinctively British qualities as the British state. Pragmatic, informal, empirical, and consensus-oriented "muddling through" was the preferred form.

Using the experience of public dissent over scientific, commercial, and state commitments to GM scientific research and agricultural innovation, we characterize these developments as constitutional in that they have reordered our understandings of science and its object "nature," and correspondingly have also tacitly reordered prevailing constitutional understandings of the agency and rights of citizens in relation to science and the state. As Jasanoff (chapter 1, this volume, 3) suggests, these emergent and incomplete reorderings are a function of "questions raised by new entities, objects, techniques, and practices that embody genetic understandings of life, but whose legal and social meanings are far from clear at the moment when scientific work first conceives of them or, through material transformations, brings them into being."

Under the terms of genetic manipulation as a sociotechnical imaginary, science's own reconceptualization of organisms was knitted, institutionally and intellectually, with a particular conceptualization of global agricultural and socioeconomic futures. As this reconceptualization gained momentum through the perceived need for commercial investment and innovation in bioknowledge, so too did the state's commitment to a corresponding vision of the citizen's agency and relations with political authority as exercised through science. This basic British constitutional settlement preceded the advent of state-commercial agricultural biotechnology but was also elaborated significantly by its biotechnical-biosocial dimensions, and by the assertion that deliberately scientifically engineered life for commercial ends was no different from "ordinary life." It was, we suggest, this stance that encouraged the normatively weighted and entrenched state scientific position that risk assessment could and did adequately encompass all relevant possible harms from GMO commercialization, with no significant unpredicted remainders. One of the most persistent effects of British government and scientific-institutional understandings of public concerns about GM as stemming from ignorance was the rigid insistence that public policy could be based only on what officials and advisers alike saw as positive scientific knowledge. That deeply rooted public concerns about the promotion of GM crops were inspired primarily by public recognition of the likelihood of unanticipated consequences, arising from scientific ignorance, was never recognized.

It is important to note that established institutional definitions of life here followed reductionist scientific accounts that had been shaped as much by commercial concerns as by scientific knowledge, in that the life of transgenic organisms, and their consequences once released into variable and complex living environments, was seen to involve no factors beyond those identified in highly restricted risk assessment protocols. Mainstream social scientific research on public attitudes, including that which claimed quasi-official status as evaluator of the government's *GM Nation?* public debate (Horlick-Jones et al. 2007), did not properly investigate this dimension of public concerns about GMOs. Research that did so, whose authors included academic members of the Unilever Contact Group (Grove-White et al. 1997; Marris et al. 2001), was not regarded unambiguously as mainstream social science, mainly because of its expressly interpretive and exploratory approach, and the critical challenges which its authors posed to UK state-scientific policy culture.

There is ample evidence that the ethical as well as risk objections that were widely recognized to be central to public opposition were inspired by a sense that official institutional culture was simply denying key issues, such as an intrinsic scientific inability to predict all significant consequences of releases of GM crops (Wynne 2001). A policy system that was supposed to represent the public interest was in effect in a state of denial (Wynne 2006), animating public disaffection and indeed fueling further public mistrust. In constitutional terms, this was a situation in which the conventional state presumption of democratically legitimated authority to define the public interest, deriving from (1) its parliamentary accountability in principle on such matters, and (2) its sovereign capacity to articulate sound scientific knowledge-authority on public policy matters, was most evidently confronted by a conflicting vision.

The GM experience illustrates the complexity and open-endedness of public attitudes in relation to issues involving science and government. It may not be too far-fetched to regard the pattern of unarticulated public issues in the British monarchist tradition as analogous to observations by anthropologists concerning subordinated populations, who work to conceal their own extensive indigenous worlds of meaning and value from their political overlords, so as to insulate themselves from annihilation and retain some autonomy and collective dignity (see, for example, Scott 1985). It has certainly been a marked problem in the UK conduct of public issues involving science and technology that public concerns have been systematically misunderstood and emptied of substance and dismissed as exaggerated and ignorant risk concerns when they have been based in very different understandings. This persistent denial calls into question the state's capacity to recognize and respond to the ontological difference between state agents and citizens in a democratic polity.

The British state's virtually complete failure to recognize that public concerns and disagreements with the state over GM might be rooted in something other than "lack of proper knowledge" was consistent with its ritualized reliance on science as authority. That dependence rendered the state incapable of acknowledging contingency or ignorance, while giving science in effect sovereign power to define public meaning exclusively in terms of "risk." Some UK scientific institutions, such as the Royal Commission on Environmental Pollution (2008), the BBSRC, and other research council counterparts, have expressly recognized this predicament and have warned that the state is unduly exposed by its making claims about the power of scientific risk assessment that cannot be borne out in experience and practice. Indeed, the Royal Commission (2008, 8) observed that ignorance and contingency are endemic in regulatory fields, and only likely to grow: "Increasingly, it will be impossible to settle questions about the environmental and human health impacts of [such] new materials consistently and in a timely fashion using traditional risk-based regulatory frameworks" (8). The Royal Commission, however, is not an executive agency and spends no public money, but is an arm's-length advisory body that the government can and does often ignore.[6]

We conclude that UK discourse and practices seem to have generated some genuinely cosmopolitan initiatives (in the sense of recognizing and respecting difference and modestly altering self-commitments and expectations) while remaining deeply influenced by what we characterize as despotic insecurities and forces. An important influence on further developments will be the clear identification and articulation, such as is undertaken in this chapter and this volume, of the intertwined, unremarked, and often implicit and indirect factors in play. Throughout the GM story, we see halting attempts to break out, by intelligent state and (in the case of Unilever) nonstate agents, aided and unevenly abetted by citizens and scientists. But opposing these moves toward opening up are a political economic dimension, as well as a UK political cultural dimension, that also have to be taken into account. Both tend in the direction of presumptive closure and hegemonic exclusion of difference and otherness and thus undermine the authentic democratic and cosmopolitan conditions that science ideally supports, and needs.

Notes

1. Institutionalized public engagement has also been criticized as something of a false-pretense: "hitting the notes, but missing the music," as Wynne (2006, 211) described it.

2. Three studies were commissioned by Unilever and overseen by the Contact Group. Two of these on public attitudes (Grove-White et al. 1997 and Grove-White et al. 2000) were co-led by one of us (Wynne). One examined expert views on the different attributes, including risks, of GM crops (Stirling and Mayer 1999).

3. One of the authors of this chapter, Brian Wynne, was a specialist adviser to the House of Lords Select Committee for the 2000 report. He thus speaks from self-

reflective direct experience. In the specific context, it would have been unrealistic to expect a full exploration of what proper dialog between science and its various publics should entail. It is also correct to say that Wynne himself could not at that time have articulated as clearly as now what further demands such two-way dialog requirements implied.

4. Thus the 2004 BBSRC crop science review repeatedly notes the need for social acceptability of chosen R&D trajectories.

5. The UK Royal Society (2009) report on *Science and Sustainable Intensification of Global Agriculture* also reflects in more authoritative form some significant dimensions of the shifted sociotechnical imaginary of the BBSRC's 2005 revision of its Crop Science Strategy. For example, its reference to public dialog needs emphasizes that "stakeholders and members of the public need to be engaged in dialogue about new research and technology options. *This dialogue should start with the problem that needs to be addressed, i.e. food security, rather than presupposing any particular solutions*" (51; emphasis added).

6. In July 2010 the United Kingdom's recently elected Conservative and Liberal Democrat coalition government announced that the Royal Commission on Environmental Pollution would be abolished in effect from the end of March 2011.

References

AEBC. 2001. *Crops on trial*. London: Agricultural and Environmental Biotechnology Commission.

Arendt, Hannah. 2005. *The promise of politics*, ed. Jerome Kohn. New York: Schocken Books.

BBSRC. 2004. *Review of BBSRC-funded research relevant to crop science: A report for BBSRC council, April 2004*. Swindon, UK: Biotechnology and Bioscience Research Council.

Burke, Derek. 2004. GM food and crops: What went wrong in the UK? *European Molecular Biology Organisation Reports* 5:432–436.

Doubleday, Robert. 2004. Institutionalising NGO dialogue at Unilever: Framing the public as "consumer-citizens." *Science & Public Policy* 31 (2): 117–126.

Ezrahi, Yaron. 1990. *The descent of Icarus: Science and the transformation of contemporary democracy*. Cambridge, Mass.: Harvard University Press.

Grove-White, Robin. 1997. Science, trust and social change. In *Science, policy and risk: A discussion meeting held at the Royal Society on Tuesday, 18 March 1997*, 53–58. Science in Society (1). London: Royal Society.

Grove-White, Robin. 2001. New wine, old bottles: Personal reflections on the new biotechnology commissions. *Political Quarterly* 72 (4): 466–472.

Grove-White, Robin, Phil Macnaghten, Sue Mayer, and Brian Wynne. 1997. *Uncertain world: Genetically modified organisms, food, and public attitudes in Britain*. Lancaster: Centre for the Study of Environmental Change.

Grove-White, Robin, Phil Macnaghten, and Brian Wynne. 2000. *Wising up: The public and new technologies.* Lancaster: Centre for the Study of Environmental Change.

Horlick-Jones, Tom, John Walls, Gene Rowe, Nick Pidgeon, Wouter Poortinga, Graham Murdock, and Tim O'Riordan. 2007. *The GM debate: Risk, politics and public engagement.* London: Routledge.

House of Lords Select Committee on Science and Technology. 2000. *Third report: Science and society.* House of Lords papers 1999–00, 38 HL. London: The Stationery Office.

Iles, Alastair. 2004. Making seafood sustainable: Merging consumption and citizenship in the United States. *Science & Public Policy* 31 (2): 127–138.

Irwin, Alan. 2006. The politics of talk. *Social Studies of Science* 36 (2): 299–320.

Jasanoff, Sheila. 2005. *Designs on nature: Science and democracy in Europe and the United States.* Princeton: Princeton University Press.

Jasanoff, Sheila, ed. 2004. *States of knowledge: The Co-production of science and social order.* London: Routledge.

Jasanoff, Sheila, and Sang-Hyun Kim. 2009. Containing the atom: Sociotechnical imaginaries and nuclear power in the United States and South Korea. *Minerva* 47 (2): 119–146.

Kamara, Mercy. 2009. The typology of the game that American, British, and Danish crop and plant scientists play. *Minerva.*

Lezaun, Javier, and Linda Soneryd. 2007. Consulting citizens: technologies of elicitation and the mobility of publics. *Public Understanding of Science* 16:279–297.

Marris, Claire, Brian Wynne, P. Simmons, and Sue Weldon. 2001. *Public perceptions of agricultural biotechnologies in Europe.* Final report of the PABE research project. Lancaster, UK: Lancaster University.

Phillips, Lord Nicholas of Worth Matravers. 2000. *The Inquiry into BSE and Variant CJD in the United Kingdom.* <http://collections.europarchive.org/tna/20090505194948/http:/bseinquiry.gov.uk>.

Royal Commission on Environmental Pollution (RCEP). 2008. *27th report: Novel materials in the environment: The case of nanotechnology.* Norwich, UK: The Stationery Office.

Royal Society. 1985. *The public understanding of science.* London: Royal Society.

Royal Society. 2009. *Reaping the benefits: Science and the sustainable intensification of global agriculture.* London: Royal Society.

Royal Society and Royal Academy of Engineering. 2004. *Nanoscience and nanotechnologies: Opportunities and uncertainties.* London: Royal Society and Royal Academy of Engineering.

Sainsbury, Lord. 2004. Quoted in "When disenchantment leads to disengagement." *Research Fortnight,* January 26, 2005.

Schofield, Geraldine. 2002. Interview with Robert Doubleday, October 8.

Scott, James C. 1985. *Weapons of the weak: Everyday forms of peasant resistance.* New Haven, Conn.: Yale University Press.

Stengel, Katrina, Jane Taylor, Claire Waterton, and Brian Wynne. 2009. Plant sciences and the public good. *Science, Technology & Human Values* 34 (3): 289–312.

Stirling, Andrew, and Sue Mayer. 1999. *Rethinking risk: A pilot of multi-criteria mapping of a genetically modified crop in agricultural systems in the UK.* Brighton: SPRU.

van Dooren, Thom. 2009. Banking seed: Use and value in the conservation of agricultural diversity. *Science as Culture* 18 (4): 373–395.

van Lente, Harro. 1993. *Promising technology: The dynamics of expectations in technological developments.* PhD thesis. Enschede, The Netherlands: University of Twente.

van Zwanenberg, Patrick, and Erik Millstone. 2005. *BSE: Risk, science, and governance.* Oxford: Oxford University Press.

Webster, Andrew. 1989. The privatization of public research: The case of the plant breeding institute. *Science & Public Policy* 16:224–232.

Wynne, Brian. 1991. Knowledges in context. *Science, Technology & Human Values* 16 (1): 1–21.

Wynne, Brian. 1992. Misunderstood misunderstandings: Social identities and the public uptake of science. *Public Understanding of Science* 1 (3): 281–304.

Wynne, Brian. 1993. Public uptake of science: A case for institutional reflexivity. *Public Understanding of Science* 2:321–337.

Wynne, Brian. 2001. Creating public alienation: Expert cultures of risk and ethics on GMOs. *Science as Culture* 10 (4): 445–481.

Wynne, Brian. 2006. Public engagement as a means of restoring public trust in science: Hitting the notes, but missing the music? *Community Genetics* 9:211–220.

Wynne, Brian. 2011. Introduction: a retrospect. In *Rationality and Ritual: the Windscale Inquiry and Nuclear Decisions in Britain.* Chalfont St. Giles, UK: British Society for History of Science; republished London: Earthscan.

12

Representing Europe with the Precautionary Principle

Jim Dratwa

The European Communities states that certain GMOs present potential threats to human health and the environment . . . [which] justifies the assessment of risks on a case-by-case basis and special measures of protection based on the precautionary principle.

The European Communities asserts that the precautionary principle has by now become a fully-fledged and general principle of international law. . . . More recently, in the specific field of GMOs, the *Biosafety Protocol* has confirmed the key function of the precautionary principle in the decision to restrict or prohibit imports of GMOs in the face of scientific uncertainty.

The European Communities further points out that in many countries approval systems are based on the need to take precautionary action. . . .

The United States argues that the European Communities has not identified how a "precautionary principle" would be of relevance.[1]

Becoming the Precautionary Principle

The text in this chapter's epigraph displays the precautionary principle as a taken-for-granted policy instrument for dealing with the scientific uncertainties surrounding the regulation of biotechnology. It associates the principle with European governance and highlights U.S. skepticism toward its application. But what is the precautionary principle and how did it come to represent Europe?

In this chapter, I trace the formulation of the precautionary principle as a critical hinge in the governance of life sciences and technologies in Europe. I aim to show that "precaution" plays a constitutional role in two respects: as a means of legitimating regulation affecting human lives (biopolitics) and as a means of legitimating European (supranational) institutions in relation to those of the member states. My inquiry pertains to the development of the precautionary principle as a construct for the regulation of technological risks and innovations, as well as the integration of scientific expertise into political decision making. At the same time, I probe the practices of representation and accountability that were drawn into the elaboration of the precautionary principle in the European Parliament.

I argue that the very idea of a "precautionary" Europe rests on a particular set of understandings about uncertainty and governance. The elaboration of the precautionary principle has been a site for working out different imaginaries of European (risk) governance, for the associated contestation between expert-based and politically grounded authority, and for the establishment of the European Parliament, the newest of the major EU institutions, as bearing particular rights and responsibilities in relation to an emerging European polity and public. Indeed, I will show that those "big," effectively constitutional settlements can be traced through "little" struggles in documents and discourse. The chapter can be read as a chronicle of the drafting and revision of a single document, the *European Parliament Resolution on the Commission Communication on the Precautionary Principle*, through the trials and tribulations of parliamentary deliberations.[2] However, precisely because those processes opened up the normative premises of the precautionary principle, the chapter does more than provide a legislative microhistory. It engages with the rights and responsibilities of decision makers, with the nature of European citizenship and representation, and indeed with the ways in which discussions of what is at risk or in need of precaution help define both an "at-risk" European public and a regime of transnational governance responsible for that public's problems.

Let us set the stage. The European Parliament receives the European Commission's Communication on the precautionary principle in February 2000, shortly after its release.[3] At the next Conference of Committee Chairpersons, it is handed over to the parliament's ENVI Committee, the Committee on the Environment, Public Health and Consumer Policy. It must then be decided which Member of the European Parliament (MEP) will actually be asked to draft it. The bargaining is tough: although this is not a legislative report, in that it will not turn into law, the precautionary principle is considered a high-profile issue by the French and German delegations. The dossier is acquired by the PSE group (the Parliament's Party of European Socialists), and eventually a French MEP, Béatrice Patrie, secures the role of primary draftsperson (*rapporteur*). Unsurprisingly, it is April 2000 before Patrie even begins work.

Explanatory Statement

As soon as Béatrice Patrie, who has never authored such a report, is appointed as *rapporteur*, documents start pouring into her office from consumer and environmental NGOs, industry, and other international and national organizations. A first discussion on "Consumer, Health, and Environmental Protection: The Precautionary Principle" takes place in the Environment Committee on July 10, 2000. It is an "exchange of views without a document," a time for exploratory deliberations allowing Patrie to lay out her preliminary vision of the issue for her peers and for them to respond briefly.

Having explained in French that she "will not delve into the substance as we do not have a document today," Patrie proceeds to "highlight the stakes" through a series of bullet points (the quotes here are from her address). She commends the European Commission for aiming to "find a common approach to the precautionary principle, an accord on the way to apprehend risks—risks which science is not in a position to evaluate." Science, even in this initial positioning, is subordinated to the precautionary principle. The stakes belong, first and foremost, to two sets of actors, Patrie says: (1) industry ("les industriels"), "and we should be careful not to stifle scientific and technological innovation"; (2) "public opinion, the European citizens who have legitimate concerns and expect legitimate decisions from European public authorities. They are counting on us to ensure a high level of protection for health and the environment." Thus, we already see citizens' concerns invoked to buttress the EU's decisions—indeed the EU authorities themselves—as legitimate.

The issues of legitimacy and of citizenship (or democracy) beyond the state are particularly salient for the EU, and we will return to them later in this chapter. We should note, however, that Patrie's remarks are closely tied to the *Exposé Des Motifs*, the tentative Explanatory Statement accompanying the Draft Report, which was already drafted at that stage. It is on her desk, together with other notes, as she addresses the committee. What follows is the official draft translation in English of the French original (my emphasis):

On 2 February 2000 the Commission published its communication on the precautionary principle.

This report sets out the European Parliament's point of view on this document. In broader terms, it expresses Parliament's views on the conditions under which the precautionary principle should be used, the criteria governing its implementation, its legal scope and the use that should be made of it at international level.

It is no surprise that various European Union bodies have shown an interest in this matter. Despite the fact that only a few years ago the precautionary principle or approach was unknown except to specialists in environmental issues, it has now taken the political world by storm ["*envahi le champ politique*"].

What some have described as the "fashion" for the precautionary principle, actually reflects the *legitimate concern* felt by public opinion, which has become extremely sensitive to such matters as a result of the recent food scandals or other disasters such as the case of the contaminated blood affair. The interest shown in the precautionary principle also reflects the *crisis of public confidence* in science, which is now recognised as not being infallible, and, more generally, in political decision makers, who are suspected of connivance with certain industrial pressure groups or simply of culpable irresponsibility.

The public no longer want[s] to bear the brunt of hazardous technological innovations. . . . They refuse to be "guinea pigs" for progress. . . .

In this respect, *public demands echo the objective set by the European Union* of ensuring a high level of protection for health and the environment.

Throughout this text, the European Parliament is represented as the voice of a European citizenry. Not only is it positioned as echoing what that public wants, but it also harmonizes "public demands" and "the objective of the European Union."

We witness here the concurrent production of a European public (at risk, recalcitrant, with legitimate concerns), European institutions (protective, precautionary, and thereby legitimate), and a European political identity linking the two. Other studies have documented how the precautionary principle has been used to forge a particular vision of Europe at the European Environmental Agency (Waterton and Wynne 2004) and at the Codex Alimentarius (Dratwa 2004). At stake here is the right of European public institutions to represent citizens. The production of a population "at risk," with *legitimate concerns*, is a prerequisite to exercising precaution through *legitimate decisions*. And what are the risks against which precaution should be taken? Risks to the environment, to health, to life, or to citizens' bodies. It is important to underscore the ontological work being done here: we see public institutions producing their own raison d'être, their own ontology as agents in the biopolitical domain.

Patrie, as draftswoman, wishes to represent the European Parliament, to garner the support of her peers, and to prompt their take on the matter as a condition for the adoption of her report. Her writing "expresses Parliament's views"—hers is "the European Parliament's point of view." At the same time, she is representing the European public (not only the handful of French voters who have deputed her), with its presumed wants and concerns, to her parliamentary peers. She is a representative of the European Parliament, but she is also representative *to* the European Parliament. Indeed, all along, she is making representations, visions of the world—*Weltanschauungen* as well as *cosmograms* (Tresch 2005; Latour 1999)—while she is invoking or abjuring particular configurations (and constitutions) of the collective, drawing and sorting possible worlds. We see in effect different aspects of the European citizenry—of the public in relation to public authorities—being constituted through the parliamentary process. We will shortly return to that process, but first drawing on John Dewey as well as on EU constitution making, we make a brief detour in search of a European public and a state.

The Problem of Discovering Europe

The European Union, and more particularly the European Parliament, embodies an experiment in internationalizing democracy. In 1953, a committee appointed by the European Coal and Steel Community wrote a draft constitution for Europe that envisaged a tripartite structure: a Council representing member states, a European Parliament representing national parliaments, and an Assembly representing European peoples directly. The European Economic Community as finally created retained only the first and the second, but two decades later proceeded to do away with the second (a parliament of parliaments) in favor of the third (a directly elected assembly). Another two decades later, in 2002–2003, the Convention on the Future of Europe revisited both the role of national parliaments and the balance between the Council and the European Parliament. At stake was the balance between people-based and state-based representation, functional and territorial representation, and direct and indirect representation. Also, perhaps because the European Parliament appears to be the main source of citizen-based Europe-wide legitimacy (as the heart pumping this scarce and delicate resource throughout the EU institutional system), issues with ramifications across national boundaries—such as issues of environmental or technological risk—acquire acute relevance here.

The legitimacy problem of European institutions, including the European Parliament, can be fruitfully related to the distinction between public and private advanced by the American pragmatist John Dewey. In *The Public and Its Problems*, Dewey offered insights into the political significance of "unanticipated consequences," which is also a defining feature of Ulrich Beck's (1992) "risk society": "The line between private and public is to be drawn on the basis of the extent and scope of the consequences of acts which are so important as to need control, whether by inhibition or by promotion" (Dewey 1927, 15). Dewey's thinking about indirect consequences pertains to the *externalities* of associated action: "all modes of associated behavior may have extensive and enduring consequences which involve others beyond those directly engaged in them" (27). Those consequences motivate Dewey's search for the state: "When indirect consequences are recognized and there is effort to regulate them, something having the traits of a state comes into existence" (12). Correlatively, a public also emerges: "Consequences have to be taken care of, looked out for . . . [T]he essence of the consequences which call a public into being is the fact that they expand beyond those directly engaged in producing them" (27). In Dewey's conception, then, looking out and caring for the indirect consequences that involve—and bring into existence—the public is the vocation of the state. Unlike Walter Lippmann (1925), who dismissed the public as a phantom, Dewey saw the public as merely "in eclipse." Like Lippmann, Dewey diagnosed a multiplicity of publics attuned to a multiplicity of controversies or problems. In examining the causes of the public's eclipse, Dewey found technologies to be the main culprit—citing, for example, the motor car as drawing peoples' attention away from politics—as technologies constitute topics of discussion and engagement that rival and surpass political matters. But today, as we have seen, technological controversies *are* political matters, and their consequences, both intended and feared, are the very stuff through which the public and the state come into being.

Dewey's conception of the public and the state (and the inquiry into "what the public may be, what the officials are" and "the problem of discovering the state") is a transformative horizon. These are not to be found lightly, and must rather "always be rediscovered" (Dewey 1927, 33). Yet something has been found in European experience which resonates with the Deweyan endeavor: this *state of precaution* whose emergence I trace, articulated in the European Parliament—looking out for and taking care of unintended consequences, in the form of both environmental and political externalities—and the European public "at risk," which the European Parliament draws on and which emerges along with it.

Pursuing the Exchange of Views

In the Explanatory Statement, Patrie highlights the "concern felt by public opinion" as well as the "concern of those in the economic, industrial and commercial sectors." In the meeting room, too, she continues to make her case (my translation): "The precautionary principle is a point of equilibrium between a guarantee of legal certainty and a guarantee of a high level of protection of the environment and of health," she says. "An agreement on the implementation of the precautionary principle is important in two respects: (1) for the policies of the European Commission; and (2) to give the Commission a clear mandate to defend our conception of the precautionary principle at the international level."

Once Patrie has "outlined the general picture" the floor is open for debate, and almost immediately dissonance arises:

Mr. B-M [in English]: We need a clear definition, and it is only fair that this should figure prominently in our debates. I would like Mme. Patrie to address a further difficulty on that matter: Once there is a definition, who should decide if and when that definition applies? One thing is clear, it should not be the individual member states, but rather the future European Agency on Food Safety. Or better still: once there is a definition, it should not be applied at all! [laughter]

The intervention and accompanying laughter underscore the uneasy relations between the precautionary principle and the emerging European Food Safety Authority (at that point still referred to as an agency). Both constructs are being concurrently shaped in the European Parliament—indeed, in the Environment Committee. Yet here, explicitly articulated, we have two key tensions: the first one pertains to the role of the precautionary principle as a European rallying cry and buttress of European identity or as a tool of dissent that can be wielded by every EU member state; the second tension pertains to the arms race between European and national independent scientific agencies, through which the legitimacy and autonomy of the European Commission and its relations of power with the member states are being displaced to the arena of scientific expertise.

What will happen to Patrie's draft report and to the parliamentary resolution? What will happen to her vision of the decision maker's role in the face of uncertainty, of procedures for bridging the gap between science and subjectivity, or of taking interested parties into account in the construction of political decisions? That story is related in the final section of the chapter, but there are intervening considerations that will help us make sense of it, situating precaution in relation to European constitution making.

Bioconstitutional Moves with the Precautionary Principle

According to Max Weber (1947, 1968) legitimacy derives from people's belief in legitimacy; more broadly, the concept of legitimacy aims to explain the acceptability of and adherence to both norms and normative systems. Like most students of the ways in which legitimacy is claimed and granted, Weber focused on the conditions under which the nation-state's rules and its right to rule can be seen as legitimate. We, however, are interested in the extension of legitimacy from the national to the supranational level. This leap raises two immediate conceptual problems: the absence of a centralized hierarchy above the state level; and the absence of an identifiable transnational *demos*, or people. With regard to the former, the absence of a supranational state does not imply the absence of any structures of government; it is precisely the practices and practitioners of one set of international organizations (European institutions) that we examine here. With regard to the latter, it is precisely such a public and polity that we have witnessed being conjured into being through the visions and actions of European parliamentarians such as Mrs. Patrie.

Representing Europe

Patrie, as we have seen, represents the European citizenry, not only through the intricacies of the electoral system but also through her practices, her writing, and her rhetoric. She has been selected for this role by a handful of voters in a previous legitimating transaction—as a European, a Frenchwoman, a Périgourdine, a woman, a Socialist, a lawyer, a mother, and a politician—and she *actively* represents all of these constituencies to her parliamentary peers. This active depiction embraces not only the particular electors who voted for her, but all their European peers. She is thus a representative *to* the European Parliament, but she also endeavors to represent the European Parliament to audiences in the EU and beyond.

Representation in politics is a Sisyphean undertaking of ceaseless reworking, reiteration, retrial, and resumption, as Dewey recognized when he saw states and publics as continuously "in the making." It is about publicly presenting oneself, one's positions and identities, one's interests and visions, and one's stakes again and again, as these change and as deliberations progress. Even though speaking *for* can mean speaking *instead of*, representation should also be understood in the political context as a demand with regard to those who are absent. "Where are the others?" is the ethical correlate to "Here is what I stand for!" Finally, such a conceptualization of representation points to a politics that involves judging and justifying and thus also making representations in a traditional sense, that is, making statements of facts and reasons and knowledge claims in the course of protesting or contending.

This analysis of representation is not limited to the parliamentary deliberations on the precautionary principle. However, the precautionary principle brings a crucial dimension to the configuration of European politics. As Patrie points out, although once unknown except to some legal specialists, and before it received high political acknowledgment (on the European scene but also for example in France, where it entered the constitution in 2005), the precautionary principle belonged *to the people*. This allows for further interplay between the European public and the European polity, as the precautionary principle is itself in a sense "a representative of the citizens." Patrie's insistence that the precautionary principle has "invaded the political field" can be taken almost literally. Indeed, in Patrie's home country, as in others, for several years now it has been the rallying cry of anti-GMO activists who have earned media attention by assaulting fields planted with GM crops. Another key feature of the precautionary principle is that it has come to represent the "European difference"—ever more deeply entrenched in the polity's self-image, particularly vis-à-vis the United States—in all risk issues from nanotechnology to climate change and from hormone beef to mad cow disease. Further, it is a notion that is specific to biopolitics, to life itself as a subject of governance.

As a soft conceptual construct or an instrument of political representation, the precautionary principle stands in a singular position at the confluence of the European public, the European polity, and life as a subject of governance. And soft it is. The work of delineating—or hardening—the precautionary principle is different from the EU's classic regulatory activity. The text that comes out of this work is weaker or softer in legal terms than directives and protocols. Yet it is all the stronger and more compelling as it leads out of the EU's normal mode of regulation-production. Regulating and harmonizing the procedures for GMO approval and release undoubtedly has a constitutional timbre in the EU; these regulations help to "make" Europe (as a single market and as a polity). But the precautionary principle, however soft, has something more: an "ought" implication,[4] with the characteristics of a moral imperative.

Tracing European Constitutionalism

Constitutional devices, simply put, are those instruments that shape the relations between state and citizens. Now what if rights and ontologies—including the status of the human, and of life itself—are jointly opened up and reconstructed? Such practices (made more common by innovations in the life sciences and technologies) and such scholarly reconceptualizations (see Jasanoff, chapter 1, this volume) place us in the domain of bioconstitutionalism. But what if the status of the state itself—together with notions of sovereignty and legitimacy—is opened up and reconstructed? This is the domain of supranational constitutionalism. The elaboration of the precautionary principle in the EU brings together both the bioconstitutional and the supranational moves.

Constitutionalism in the EU setting encompasses five salient features. First, there is the "direct effect" or direct enforceability of constitutions.

Constitutional norms confer rights. For example, the EC treaties are *constitutional*, in that these legal agreements among sovereign states give rise to new rights enforceable even by individuals and even against member states and national law.

Second, constitutional law is generally regarded as higher law, trumping other levels of law. This has given rise to interesting quandaries in the EU, pitting EU law against member states' constitutions (which count as higher law only *within* those nations).[5] The hierarchical principle also means that, as applied to international organizations such as the WTO or the Codex Alimentarius, a constitutionalizing discourse can serve to deflect criticism of or dissent from the work of these bodies by conferring an untouchable constitutional quality on their exercise of power.

Third, constitutionalism at the supranational level cannot avoid being marked by the forms, structures, and creation processes of existing national constitutions (and it is also marked by the evolving substance of those constitutions, hence resonating with what might be called issues of constitutional significance, such as legitimacy, sovereignty, subsidiarity, human rights, and, more recently, environmental protection). This includes not only the founding myths and the "we-identity" constitutive of a demos (Weiler et al. 1995, 5; Scharpf 1998, 1999), but also the very desire for a formal constitution. Indeed, a written constitution is itself a token, a thing of sentimental value, a symbol of nationhood, and thus part and parcel of the founding myths and we-identity of the polity it governs (see Jasanoff 2003b).

Fourth, constitutionalizing a political order also reifies it, to some degree. By characterizing a polity as constitutional, one transforms its complexity and messiness—its mundane goings-on, the bric-a-brac of its component parts, the bargains ceaselessly struck among its diverse rules and principles—into an ordered, unitary, more elevated whole. Not that this puts an end to the heterogeneity below, but, as noted by Robert Howse and Kalypso Nicolaidis (2001, 228) in regard to the WTO, "Individual elements become less easily contestable. The WTO becomes reified as something one is either for or against."

Fifth, to constitutionalize is to strive for a distribution of competences, as exemplified in the EU by debates over the desirability of a clear catalog of competences.[6] Such a division of labor requires the balancing of competing values or the hierarchical ordering of competing organizing principles. This can mean placing law, or the rule of law, above politics. It can mean, in the case of constitutionalism at the WTO, for example, placing free trade above social and environmental concerns. And it can mean, in the context of the EU, placing some concerns about human health, safety, and the environment above trade, albeit not above the single market.

From this perspective, the precautionary principle may be seen as an instrument for the organization of competing values. In addition to its role in embodying a European polity, and communicating a European identity, and in addition to its other distinctive traits discussed previously (appealing to grassroots, biopolitics, and a moral legitimacy beyond mere legality), the capacity of the precautionary principle to adjudicate among competing norms underscores its constitutional dimension (see also Rip 1999). In this chapter, we turn to the role of the precautionary principle in supranational constitutionalism, specifically in shaping rights and relations among different actors involved in the EU's multilevel governance.

The Parliament's Precautionary Principle

The European Commission, the EU's highest executive authority, is present at the European Parliament in several ways. Formally, most of the texts that the European Parliament responds to originate in the Commission. Further, Commission representatives accompany these texts and are present in both committee and plenary. But there is yet another mode of presence, less conspicuous though no less important. Texts can be particularly sensitive and Commission officials particularly zealous in attending to them. In our case an official who participated in drafting the Commission's text—the Communication on the Precautionary Principle—also intervened in drafting the European Parliament's response. This shed a particular light on the contrasts in thinking between Commission and the European Parliament that merit our attention.

The European Commission and the European Parliament
It is instructive to review in detail an annotated paragraph that the Commission official in question submitted to Patrie on seeing the draft report (as in the original, the Commission official's additions are underlined and the unwanted initial text crossed out):

13. The European Parliament recognises that every political decision-maker (of the Community?) To whom is this addressed? If it is to third States difficulties exist (extra-territorial application); if member States are included, problems for the internal market. and risk manager is entitled to determine ~~the level of protection required, in accordance with the following principles:~~

1) ~~the chosen level may not be lower than that laid down by European or international standards~~ too vague and imprecise

2) protection of human health and <u>protection of</u> the environment must take precedence over all other considerations

3) there is no minimum risk threshold below which the precautionary principle may not be used

4) it is not necessary to demonstrate <u>conclusively</u> a cause and effect link between the phenomenon or product generating the potential risk and the feared effects

The annotations in the first paragraph focus on the sovereignty of the decision maker, and more specifically on the distribution of sovereignty among the actors who together compose the decision-making authority. Does the right to determine the level of protection belong to the European Community, to its member states, or to states beyond the EU? At stake here is the stabilization of the precautionary principle not only as a matter of international law, but also as an *internal* matter, that is, in the settlement in progress among EU institutions. Simply put, the Commission wants the European Parliament to adhere to its views. In fact, as I will further show, internal action is understood as a "stepping stone" to the international domain.

Trials and Tribulations through Amendments

While Patrie was presenting her Draft Report in the ENVI Committee, another MEP, Evelyne Gebhardt, was drafting and discussing her own separate Opinion on the same matter in the JURI Committee (on Legal Affairs and the Internal Market). Those deliberations took place in the same week in mid-September 2000. I will recount two episodes here, one in each committee, to show how the precautionary principle operated as a motif of regulatory reflexivity and of elicitation of a European (regulatory) identity geared toward both the internal and the international scene. An intervention by Gebhardt opens the proceedings at the JURI Committee:

Mrs. Gebhardt [in German, with simultaneous translation in English]: Thank you, Mr. Chairman. When I took up this "precautionary principle" issue I thought it would be interesting, but it was actually even more fascinating than I had thought. . . . The precautionary principle features, as you know, in article 174(2) of the Treaty, but also in international law, and is also a matter of contention at the WTO, and also at the Codex Alimentarius—with the work there on General Principles, which will lead next year in that framework to a definition of the precautionary principle. That is also why it is important that we in the Parliament should provide some input and backing.

As an example of how it works, and also showing that we have it in our EU legislation, we can look at legislation on GMOs—Directive 90/220 and its revision. It is designed to avoid any risks, and Article 22 contains a safeguard clause which authorises the member states to impose restrictions or bans on a temporary basis should new or additional information emerge providing evidence of a danger to the environment or public health. The precautionary principle applies to internal legislation—it would be good if it was applied outside too.

This comment underscores a key aspect of the precautionary principle: not only does it rely and build upon previous articulations, but it is also always open to further revision, even if (as in the glancing reference to the Codex Alimentarius) those prospects subsequently fail to come to fruition. Here again we see Dewey's pragmatist state always in the making: in the repeated adaptation of the principle in different contexts, as well as in its substantive provisions on temporary measures destined to be revisited in the light of new circumstances. Further, Mrs. Gebhardt's deploring that "In the [European Commission's] communication, the precautionary principle is described as a variant of risk management. It comes into effect only if a risk assessment has already been carried out and no clear scientific evidence of any danger has been brought to light" and her assertion that "the precautionary principle must be applied proactively and not just at a stage when doubts have arisen" can be read as a refusal to have the principle confined to doubts that have already passed through risk assessment, or even as a refusal of the risk analysis framework itself, with its claims to scientific rigor and universal applicability. This implication is immediately challenged:

Mr. M [in English]   In paragraph 4 of the conclusions, you [Mrs. Gebhardt] argue that the precautionary principle should be applied "proactively." It is in line with what you said now about "avoiding any risk." That is somewhat problematic. That would make it a principle hostile to innovation. There is always a risk with innovation, but we should still innovate.

The Chair, Mr. V. [in Dutch with simultaneous translation in English]   Perhaps we should not be excessively subtle. Reasonable suspicion itself can actually be very subjective. Life is full of risk from birth to death, life itself is one long risk! But perhaps we should not get bogged down. [laughter]

The Chairman's mention of reasonable suspicion refers to another statement in Gebhardt's draft Opinion: "Consideration must be given to the issue of the measures which can be taken in accordance with the precautionary principle if the risk assessment consists solely of *justified suspicions* held by experienced practitioners in the field concerned and no scientific proof is available" [emphasis added]. Charitably read, this proposal could be seen as an institutionalization of the precautionary principle at its best. In this reading, precaution would come into play when "experienced practitioners" agree that there is no scientific proof but also that a reasonable suspicion of risk exists in their collective judgment. Precaution in these circumstances would entail a loosening of the yoke of presumed scientific certainty and lead to a potential broadening of the expert base to include lay perspectives. Furthermore, suspicions would have to be *justified*, that is, shored up by claims that can be compared and contradicted. It would thus become an exercise in public reason, not scientific fact finding.

We now turn to the Environment Committee, of which a meeting is underway; Patrie is busy laying out her draft report and her own views on the precautionary principle:

Mrs. Patrie [in French]   If we come to an agreement within the European Union—Commission, Parliament, and Council—we will have a common European conception. Moreover, the Commission could implement it in all Community policies, not only the environment. And also in international arenas!

This is a particularly forceful statement of the notion of incremental leveraging, where one agreement leads to the fulfillment of further common actions (Dratwa 2004). A possible corollary—indeed Patrie's hope—is that the precautionary principle which is thus agreed will be homogeneous, transferable to other polities, national and international, and to other policy domains. This is consistent with the *one for all* stance—one precautionary principle for all policy domains—that will ultimately be adopted by Patrie and by the European Parliament.

Internal action is understood as a stepping stone to the international domain, yet interactions among EU actors are not just stepping stones. In this process, connections are tied and tightened; ideas, texts, and policies are exchanged, fine-tuned, or coordinated; out of these goings-on, it is a European front that emerges. It is a front in all its meanings, spatial and social: a defining line or boundary; the part that is nearest to the normal viewer; an area of activity involving effort; a group of actors with something in common who try together to achieve certain shared goals. This is what we see here and what we saw earlier with the Commission official's intervention as well as with Gebhardt's reference to Codex and her appeal to "backing" the European mobilization for precaution:

Mrs. Patrie continues   As in the debate on the food authority this morning, it is appropriate here to distinguish between risk assessment and risk management, which the Commission's Communication does very well, that is by returning to everyone their true responsibilities. . . . What is lacking in the Commission's Communication, however, is the right of the political decision-maker to determine the desired level of protection—provided one does not contravene a series of criteria. In that regard, no "zero risk," of course, but no risk "too small for measures to be taken" either. As to cost-benefit analysis: it is not only economic factors that it should take into account but also the sociological, human, more subjective ones.

Here Patrie touches on the links between political and technical: on the technical side, she refers to the role of risk assessment and risk management, of course, but she also calls attention to the rights and responsibilities of the decision maker, the discretion to act even on low risk, and the need to include factors with subjective aspects. On that last point, it is instructive to turn to the final Explanatory Statement of the Draft Report (official translation of the French original drafted by Patrie):

It would be useful for studies that examine factors of this kind [i.e. contributions from humanities and the social sciences along with usual scientific expertise] to be made available to public decision-makers. In fact, although decision-makers in all fields enjoy powers of discretion bordering on the subjective, it is vital in a democratic decision-making process that the criteria on which decisions are based should not be arbitrary and can be explained as openly as possible to the public.

Patrie recognizes that decision making necessarily comprises elements of subjectivity. Rather than deny discretion, however, or cover it up with the implied certainty of science-based decision making, she addresses the question of how to prevent subjectivity from turning into arbitrariness. To that end, the factors on which decisions are based should be made explicit and then also explained. Similarly, the decision should be justified on the basis of the factors that experts took into account. Further, because no single factor brings certainty, it would be desirable to invoke a diversity of factors. The consideration of other factors is crucial and connects with language adopted at the Codex Alimentarius, with the talk of "other legitimate factors" (for factors *other* than narrow scientific expertise, with which to *legitimate* decisions, see Dratwa 2002).

We see in these examples of textual fine-tuning how the traditional risk assessment/risk management (or science/politics) dichotomy is subtly reformulated into a discourse on uncertainty and subjectivity. On the one hand, this shift takes the debate away from exclusive control by experts into the domain of representation and public justification. On the other hand, it buttresses the precautionary principle itself as a legitimating concept. It allows the EU polity to move from a more bureaucratic-rationalistic mode of self-legitimation (based on technical risk assessment) to one that foregrounds the politics of uncertainty and positions the institutions of the EU—as opposed to those of the member states—as the ones best equipped to govern those uncertainties. Close textual analysis shows how an understanding of what is at risk, and who should be protecting people against such risks, gets negotiated during committee proceedings. This demonstrates how political bodies, through their drafting procedures, mediate in effect among multiple conceptions of European-ness, as well as among different scientific objectivities, as embodied in the independent scientific agencies of the member states and the EU itself. But much work still remains for Patrie and her parliamentary peers with regard to the distribution of sovereignty between the member states and EU institutions.

Resolution

Various amendments (10 in total) were tabled on the additional paragraphs of article 13 of the Draft Report. An example is the revision proposed to paragraph 2 by Jean Saint-Josse, the head of *Chasse, Pêche, Nature, et Tradition* (CPNT—Hunting, Fishing, Nature, and Tradition), a small French political party that managed to get two deputies elected to the European Parliament in 1999:

(2) protection of human health *(three words deleted)* must taken [sic] precedence over all other considerations;
Amendment 56
(original tabled in French; official translation in English)

The amendment consisted merely of deleting the words "and the environment" from the top of the hierarchy of considerations. The move was perhaps related to the CPNT's vocal antagonism toward the Green Party at the European Parliament, the French "ecological" party affiliated with the Greens, and the French Minister for the Environment, Dominique Voynet (a member of the French environmental party, and head of the European Council of Ministers of the Environment at the time). Deleting three words was a remarkably efficient constitutional move to disempower all these!

An amendment pertaining to paragraph 1 tabled by Inger Schörling and Patricia McKenna was designed not to contravene but to further specify the initial text:

(1) the chosen level may not be lower than that laid down by European or international standards, *such as limit values, WHO recommendations and environmental policy targets*;
Amendment 55
(original tabled in English; italic indicates language added or changed by the amendment)
This amendment was designed to relieve the worry about vagueness and imprecision previously expressed by the Commission official. Indeed, the proposal was later adopted as an amendment to the Draft Report and can be found in the final Resolution.

Regarding the first part of article 13, touching on the sovereign entitlements of the decision maker, we recall that the principled right to determine the level of protection was deemed misguided by the Commission official. Addressing that lack of focus, Hans Blokland tabled this amendment:

13. [The European Parliament] *Takes the view that, in common with the other members of the WTO, the Community (nine words deleted)* is entitled to determine the level of protection required, *in particular in the field of protection of the environment and of human, animal and plant health; the precautionary principle is an important policy instrument in this respect* in accordance with the following *assumptions*:
. . .
Amendment 53
(original tabled in Dutch; official translation in English)

The amendment specifies the environment and human, animal and plant health as the fields of protection. What is more, the key question of authority posed by the Commission official with regard to the language of the initial Draft Report—"To whom is this addressed ?"—finds an answer here. The entitlement does not pertain to "every political decision-maker and risk manager," as stated in the original draft, but rather to the community.

Patrie herself posed a different answer to the question of sovereignty in another amendment:

13. [The European Parliament] Recognises that the *European Community and the Member States are* entitled to determine the *desirable* level of protection, in accordance with the following principles:
. . .
Amendment 54
(original tabled in French; official translation in English)

In her formulation, it is not only the European Community but also its members that may make the relevant determination. This goes against the proposal by Mr. B-M noted previously, and it acknowledges—without resolving—the tension between the precautionary principle as a shared European construct disseminated among European states (who may if they choose implement it differently) and as a principle wielded Europe-wide solely through common decisions (i.e., through Brussels).

Now that two alternative framings of rights have been laid out, which is adopted? There are two notable flip-flops along the way. At the Environment Committee meeting of November 21, Patrie's amendment passed and Blokland's was defeated. But a month later, Blokland's proposal gets a new lease on life: after the Draft Report was adopted as the Environment Committee's Motion for a Resolution, Blokland's proposal was tabled again as an amendment to that resolution, not by Blokland himself but by his party. And on December 14, 2000, Blokland won the day after all. His proposal made its way through the plenary session and into the final document, as a majority of his parliamentary colleagues, and notably those of his own party (the European People's Party, the largest party in the European Parliament), scrupulously followed their voting instructions.

The amendment that passed, however, was slightly but significantly different from the one we saw earlier. The European Parliament's resolution was finalized as follows: "Recognizes that the European Community, in common with the other members of the WHO, is entitled to . . ." It is no longer the World *Trade* Organization that is referred to, but the World *Health* Organization. I myself thought at first that this was a typographical or translation error in the English version, but the other language versions (notably the French and the Dutch) were similarly altered. An assistant to Blokland assured me that this actually was the desired text, emphasizing that this reference to the WHO at the beginning of the article coincided with the previously amended language of paragraph 1 referring to the WHO as a source of international standards.

Such a shift of emphasis from trade (and its possible correlate, *protectionism*) to health (and thus *protection*) is consistent with a move toward biopolitical rationales for governing, in which the precautionary principle can be invoked for the protection of life. This dovetails with broader shifts in the politics of globalization, toward the construction of biopolitics at a supranational level, as the contributions of Mariachiara Tallacchini and Kaushik Sunder Rajan (chapters 8 and 9, respectively) to this volume also argue. Very specifically, here under the heading of (public) health and its protection, the precautionary principle repositions and refounds the legitimacy of the EU institutions both within Europe and in a wider international context.

After the Resolution?

What happens after the adoption of the resolution? The formal end product alone, the text finally adopted by the European Parliament in plenary session, though deserving of careful attention, is certainly not the only thing that "happens" through the intricate process discussed previously. The process does not end with the product, and it is interesting to consider the aftermath of the adoption. The implementation phase of a policy is a crucial, though neglected, aspect of policy analysis. In the case of this Resolution on a Communication, however, we have the softest of soft law, and we must look closely to find much by way of implementation.

What happens, to varying extents, is a series of movements and reconfigurations: for the precautionary principle itself, which gained not just increased leverage but a new political identity; for Béatrice Patrie, who was deemed to have done a very good job with her first report and was soon entrusted with a second one; for the Environment Committee; for the environment as a policy domain; for the Party of European Socialists (and its French delegation and group at the Environment Committee); for the European Parliament; for the other EU institutions in their relations with the European Parliament; and so on. Debates on the European Union's regulatory framework on GMOs and on the emerging European Food Safety Authority were also closely tied to the shaping of the precautionary principle at the European Parliament.

As regards the document on the precautionary principle that we have followed to this point, the resolution stipulates in its final paragraph:

38. [The European Parliament] Instructs its President to forward this resolution to the Council, the Commission, the Economic and Social Committee, the Committee of the Regions and the parliaments of the Member States.

And as stipulated in the last paragraph of the Report's Explanatory Statement:

While once again commending the quality of the Commission's work, to which the present report is intended to be a constructive contribution, there is a need to urge the Council to make known its own approach to the precautionary principle in the near future and adopt a resolution on the subject by the end of the year 2000.

This forwarding and urging underscore a distinctive feature of the collective articulations of the precautionary principle—as an end as well as a means, a completion and a stepping stone, a link in the chain.

Conclusion: Precaution as a Constitutional Principle

We have seen the precautionary principle forged at the European Parliament functions as a two-pronged constitutional appliance, enabling a rethinking of constitutionalism in relation to both biopolitics and multilevel governance. In tracing the development of this principle, we have taken official texts seriously, together with the underlying elements of administrative legal culture, in keeping with the methodological approaches opened by the introductory chapter to this volume. In fact, we have found the precautionary principle to be a device for coproduction enabling not only the concurrent production of natural and social orders—through bioconstitutionalism and supranationalism—but also the joint articulation of *is* and *ought*, of knowledge and action, of risk and predictability (and protection from risk), together with the articulation of a European public and a European public authority.

I have suggested that the precautionary principle serves as a constitutional cornerstone of the emerging European polity and its public in several senses. First and most important, it incorporates attempts to secure a unity of purpose (or a unity of principle) among EU institutions engaged in the establishment of Europe as a *state of precaution.*

The principle has additional distinctive strengths: as a representative of citizens' demands, as a symbol of European difference, as a linchpin of biopolitics and, by appealing to public values, as placing legitimacy beyond mere legality. We have also observed the constitutional role of the precautionary principle in the organization of competing values. The precautionary principle as we have followed it does not merely bring about a distribution of regulatory competences; rather, the debate on precaution elicits an explication of competences among those who grapple with it. It requires EU actors to unpack and extend *legitimacy* beyond *state-based* as well as *science-based* means of legitimation, impelling actors to elucidate which factors come into the making of decisions.

Patrie's gravest fear at the outset was "to have missed out something," to have left a relevant input, perspective, or issue out of the elaboration of the text on the precautionary principle. There was a real sense of purpose driving a number of MEPs, including Patrie. As we saw, their mandate commits many of them to take upon themselves the task of *representing the European collective*: on the one hand, they carry out the modes of political representation described previously; on the other, they engage in the enterprise of imagining, advocating, and confronting possible states of the world that they are called upon to regulate. Should protection of human health and/or the environment take precedence over other significant considerations (such as free trade)? Should the public be protected from becoming guinea pigs for industrial progress? Under what constraints should European political decision makers and risk managers be entitled to determine the required level of protection? How much should be determined through centralized action by Brussels and how much should be left to member states or specialized agencies such as Codex?

Just as the European Parliament was called into being by an emerging Europe, and just as its members are called into being by their respective publics, so the European Parliament and its members also call certain publics and certain Europes into being by acting for them. The precautionary principle, as both an instrument of social imagination and a field of action, serves in these respects a particular constitutional function.

Notes

1. World Trade Organization, February 7, 2006, Interim Reports of the Panel established by the Dispute Settlement Body in *European Communities—Measures Affecting the Approval and Marketing of Biotech Products* (WT/DS 291,292, and 293 INTERIM), §§7.77 to 7.80.

2. For a discussion of the European Commission's Communication (CEC 2000) on the precautionary principle, see Dratwa 2002. The Communication avoids defining the precautionary principle. It does concede that in situations of scientific uncertainty and risk (specifically "where preliminary objective scientific evaluation indicates that there are reasonable grounds for concern that the potentially dangerous effects on the environment, human, animal or plant health may be inconsistent with the high level of protection chosen" [CEC 2000, 2]), "A decision to take measures without waiting until all the necessary scientific knowledge is available is clearly a precaution-based approach" (CEC 2000, 7). This constitutes a useful working definition for purposes of this chapter.

3. In the EU's peculiar political architecture, three bodies—the European Commission, the European Parliament, and the Council (bringing together the member states' ministers), forming the so-called institutional triangle—act as colegislators.

The European Commission, holding the right of initiative, had originated the Communication, while the Council and the European Parliament, in the case of this nonlegislative file, were called upon to respond to it.

At the European Parliament, as we will see, the lead for the preparation of the parliament's response (its Resolution) is entrusted to a specific Committee and to a specific draftsperson. The latter must then prepare—to be adopted in that Committee—a Report comprising a Motion for a Resolution and an Explanatory Statement. In parallel, one or more other Committees adopt their own separate Opinion, which will also accompany the other documents. These procedures also give rise to several sets of amendments. Ultimately, after adoption in Committee and duly amended, the Motion for a Resolution is tabled for adoption in the European Parliament's plenary session. The end product of all this is the Resolution.

A discussion of the Commission's Communication (CEC 2000) and the precautionary principle it communicates is offered in Dratwa 2002. The Communication avoids providing a definition of the precautionary principle. It does concede, nonetheless, that in situations of scientific uncertainty and of risk (specifically "where preliminary objective scientific evaluation indicates that there are reasonable grounds for concern that the potentially dangerous effects on the environment, human, animal or plant health may be inconsistent with the high level of protection chosen" [CEC 2000, 2]), "A decision to take measures without waiting until all the necessary scientific knowledge is available is clearly a precaution-based approach" (CEC 2000, 7). This constitutes a useful working definition for the purpose of this chapter.

4. For a discussion of that peculiar ought implication, distinguishing the prudential dimension from the moral dimension of the precautionary principle, see Jasanoff 2003a.

5. In an interesting diagnosis, although not resorting to any such inventory of what it is to constitutionalize, Joseph Weiler does bring together our first and second items in this regard: "National courts are no longer at the vanguard of the 'new European legal order,' bringing the rule of law to transnational relations, and empowering, through EC law, individuals vis-à-vis Member State authority. Instead they stand at the gate and defend national constitutions against illicit encroachment from Brussels. They have received a sympathetic hearing, since they are perceived as protecting fundamental human rights as well as protecting national identity. To protect national sovereignty is passé; to protect national identity by insisting on constitutional specificity is à la mode" (2001, 60). Some aspects of this *constitutional* process—that is, constitutive in ways we are currently examining—can thus be "anti-constitutional"—that is, not in keeping with nations' constitutions or constitutional courts. It is in the former sense rather than the latter that I refer to constitutionality as constitutive of a constitutional order, whether unwritten, or not sanctioned by a formal constitution.

6. This was notably the case with the debate on "governance" in the EU at the turn of the century, as crystallized in the white paper on European governance (CEC 2001), and with the Convention on the Future of Europe, the explicitly constitutional mobilization that took place subsequently (in 2002–2003) and that was pervaded by appeals to establish a precise "who does what" that would curtail the "creeping competencies" of EU institutions seen by certain national actors as encroaching on the member states' prerogatives.

References

Beck, Ulrich. 1992. *Risk Society: Towards a New Modernity*. London: Sage.

Commission of the European Communities (CEC). 2000. *Communication from the Commission on the Precautionary Principle*. COM(2000)1 of 2 February 2000.

Commission of the European Communities (CEC). 2001. *European Governance: A White Paper*. COM(2001)428 final of 25 July 2001.

Dewey, John. 1927. *The Public and Its Problems*. Athens, Ohio: Ohio University Press.

Dratwa, J. 2002. Taking Risks with the Precautionary Principle: Food (and the Environment) for Thought at the European Commission. *Journal of Environmental Policy and Planning (special issue on Risk and Governance)* 4 (3): 197–213.

Dratwa, J. 2004. Social Learning with the Precautionary Principle at the European Commission and the Codex Alimentarius. In *Decision Making within International Organizations*, ed. B. Reinalda and B. Verbeek, 215-228. London: Routledge.

Howse, Robert, and Kalypso Nicolaidis. 2001. Legitimacy and Global Governance: Why a Constitution for the WTO Is a Step Too Far. In *Equity, Efficiency and Legitimacy: The Multilateral System at the Millennium*, ed. Roger Porter,

Pierre Sauve, Arvind Subramanian, and Americo Zampetti, 227–252. Washington, D.C.: Brookings Institution Press.

Jasanoff, Sheila. 2003a. A Living Legacy: The Precautionary Ideal in American Law. In *Precaution, Environmental Science, and Preventive Public Policy*, ed. Joel Tickner, 227–240. Washington, D.C.: Island Press.

Jasanoff, Sheila. 2003b. In a Constitutional Moment: Science and Social Order at the Millennium. In *Social Studies of Science and Technology: Looking Back, Ahead, Yearbook of the Sociology of the Sciences*, ed. Bernward Joerges and Helga Nowotny, 155–180. Dordrecht: Kluwer.

Latour, Bruno. 1999. *Pandora's Hope*. Cambridge, Mass.: Harvard University Press.

Lippmann, Walter. 1925. *The Phantom Public*. New York: Harcourt, Brace and Company.

Rip, Arie. 1999. Contributions from Social Studies of Science and Constructive Technology Assessment. ESTO study on technological risk and the management of uncertainty. Enschede, The Netherlands: University of Twente.

Scharpf, Fritz W. 1998. Interdependence and Democratic Legitimation. Working Paper 98/2. Köln: Max Planck Institut für Gesellschaftsforschung.

Scharpf, Fritz W. 1999. *Governing in Europe: Effective and Democratic?* Oxford: Oxford University Press.

Tresch, John. 2005. Cosmogram (Interview with Jean-Christophe Royoux). In *Cosmogram*, ed. Melik Ohanian and Jean-Christophe Royoux, 67–76. New York: Lukas and Sternberg.

Waterton, Claire, and Brian Wynne. 2004. In the Eye of the Hurricane: Knowledge and Political Order in the European Environment Agency. In *States of Knowledge: The Co-Production of Science and Social Order*, ed. Sheila Jasanoff, 87–108. London: Routledge.

Weber, Max. 1947. *The Theory of Social and Economic Organization*. New York: The Free Press.

Weber, Max. 1968. *Economy and Society: An Outline of Interpretive Sociology*. New York: Bedminster Press.

Weiler, Joseph H. H. 2001. Federalism without Constitutionalism: Europe's Sonderweg. In *The Federal Vision. Legitimacy and Levels of Governance in the United States and the European Union*, ed. K. Nicolaïdis and R. Howse, 54–71. Oxford: Oxford University Press.

Weiler, Joseph H. H., Ulrich R. Haltern, and Franz C. Mayer. 1995. European Democracy and Its Critique. *West European Politics* 18 (3): 4–39.

13

Conclusion

Sheila Jasanoff

Revolutionary changes in the scientific representation and technological malleability of living matter in the twentieth century entailed equally far-reaching changes in the accommodation of life, especially human life, within the legal cultures of nation states. This book's principal objective has been to trace some of the most significant shifts, from disparate but complementary perspectives and with a diversity of methods, showing how they fit within a framework that we call bioconstitutionalism. The cases recounted here represent on one level snapshots of transformations still in progress. Each chapter analyzes the responses of a particular kind of institution—hospital, ethics committee, court, scientific society, biotech company, regulatory authority, science policy agency, and even the Catholic Church—to specific biological ideas, constructs, instruments, and practices. Taken together, however, these cases point toward fundamental realignments in the legal and ethical relations between human beings, their bodies, their families and genetic kin, their relationships with other species, and their institutions and norms of government. By drawing together approaches from the history of science, science and technology studies, law and legal philosophy, bioethics, and comparative politics this interdisciplinary collection seeks to make sense of a technoscientific revolution that is at the same time also a revolution in humanity's capacity for self-understanding and self-control.

Observers of biology and society have regaled us in recent years with an array of new constructs built on the morpheme "bio," or its more ample partner "biological." Beginning with Michel Foucault's seminal concepts of biopower and biopolitics, the lexicon of the social sciences and public policy now also includes terms like biocapital, biohazard, bioidentity, bionationalism, biosafety, biosecurity, biosociality, biosocieties, and biological citizenship. Meanwhile, science and technology have been busily deploying their own creative energies, with coinages such as bioaccumulation, biofuels, bioluminescence, and bioremediation. In a terminological world that seems at times oversaturated with bioliveliness, why introduce yet another compound? What does "bioconstitutionalism" add to our understanding of the dynamics of science and law, or of life's changing social meanings? Does it amplify or elaborate on the sizeable analytic literatures that have already grown up around genetics and the life sciences? Does it contribute to conceptual clarity and responsible stewardship?

We argue that it does all this and more. The developments described in the preceding pages can be seen, in effect, as elements of a deep-going transformation in the relations between life and law, responding to the transgressive nature of the new biosciences and biotechnologies. As biology crosses conceptual lines that have long been foundational to legal thought—between life and nonlife, human and nonhuman, individual and collective, predictable and unpredictable—we see in progress a profound rethinking of the rights, duties, entitlements, and needs of living entities in relation to law and the state. A capacious concept is needed to alert us to this phenomenon and to illuminate its multiple facets, a concept moreover that allows us to look beyond the courts to other places in society where significant principled adjustments to new representations of life are taking place. Bioconstitutionalism serves that purpose. This conclusion seeks to abstract from the specific case studies some more general observations about the nature of bioconstitutionalism and its possible evolution in the twenty-first century. In so doing, the chapter also highlights the book's contributions to legal and social theory, bioethics, and political thought.

Rereading the Texts of Law

We began by stressing the mutual interpolation of the texts of law and the "texts" of DNA. More than a metaphor, the continual interweaving of legal and scientific texts suggests a need to reconsider some of the limitations of classical legal analysis, which has tended to give primacy, and hence conceptual dominion, to its own command of the written word. The methods and analytic strategies adopted in this book argue for foundational changes in how we read and interpret the law for purposes of normative analysis. In this respect, the book serves as a primer not only for students of *biology* and society but also for critical studies of *law* and society in technologically advanced democracies.

A recurrent point stressed by all the contributors is that context matters in regulating life, a point that theorists in search of pure principles tend to ignore. Thus, there is no singular way to determine whether or when a new biological construct is entitled to ethical treatment as incipient human life. The answers provided by both ethics and law depend on contingent factors but nevertheless achieve durable articulations within specific political cultures. Britain's empiricist scientific tradition, in concert with a thousand years of the common law's incremental approaches to fact finding (chapters 3, 4), draws the line between life and nonlife in different ways and with radically different scientific and legal consequences than Germany's postwar *Rechtsstaat* (chapter 3) or Italy's ancient politics of church and state (chapter 5). Different cultural approaches to elite formation, together with divergent national civic epistemologies, lead to different demands for citizenship and participatory rights at the frontiers of uncertain, potentially ungovernable technological developments (chapters 8, 10, 11, 12). Preexisting orderings of state and market seem to foreordain different constructions of human subjectivity in India and the United States (chapter 9). Even within a single nation's legal system, a context-sensitive reading of judicial opinions, alive to the influence of political culture, reveals unsuspected complexities in the definition of individual rights, and their limits, within a functioning moral community (chapters 6, 7).

These embedded histories and trajectories caution us against univocal grand narratives, such as medicalization, geneticization, and even eugenics (chapter 2). For law and ethics, a crucially important point is that values are at once more contingent—that is variable across time and space—than doomsday scenario builders suggest, and more deeply embedded in institutional cultures and practices than pragmatic reformers or technology enthusiasts concede in their calculus of risks and benefits at the frontiers of scientific discovery.

Biopower and Bioconstitutionalism

As this book's authors take pains to point out, the notion of bioconstitutionalism probes state-society relations at a level of concreteness that is often missing from discussions of biopower. Bioconstitutionalism directs us first and foremost to the position of the individual human within a constitutional order that recognizes liberties, grants rights, and provides means of representation and redress when those rights and liberties are at stake. Bioconstitutionalism, as elucidated throughout this collection, allows us to look determinedly in two directions. On the one hand, it is a lens through which we can examine the state's role in deciding when new entities, new subjects, or new rights will be certified as deserving recognition: for example, when the Italian state converts IVF embryos into de facto citizens, or the U.S. judiciary deliberates whether felons deserve more or less informational privacy than innocent citizens. On the other hand, the framework of bioconstitutionalism also displays the power of human subjects to articulate new claims vis-à-vis governing institutions, thereby demonstrating the productivity of constitutional ideas as resources for bottom-up self-fashioning. Somewhat counterintuitively, and cutting against deterministic predictions, biological knowledge serves in this sense as an enabling instrument for democratic constitutionalism.

Linking biological developments to constitutional thought necessarily opens up comparative dimensions that often get sidelined in the grand narrative of biopower. A major achievement of this book is to show, through comparisons within as well as between chapters, how the classification and ordering of biological entities and processes are mediated through different constitutional notions of the state's responsibility toward its citizen-subjects. Thus, my chapter describes how a U.S. ethics committee replays and reinscribes American political commitments to pluralism and liberal individualism. Coming from a different angle, that of U.S. constitutional jurisprudence, David Winickoff shows how federal courts base the analysis of prisoners' rights on tacit models of risk (to society) and responsibility (of the information-seeking state). Kaushik Sunder Rajan connects the different approaches toward the experimental subject in Indian and U.S. pharmaceutical development to fundamentally different models of state action and state obligation for creating public goods. Robert Doubleday and Brian Wynne describe a tug of war between ancient monarchical traditions and newer admissions of uncertainty and limits in Britain's on-again, off-again attempts to involve the public in managing agricultural biotechnology. And in Mariachiara Tallacchini's cross-national comparison of xenotransplant regulation and Jim Dratwa's account of Europe's precautionary principle, one sees the nascent European suprastate attempting to define its custodial relations with its diverse populations, independently of their concurrent national identities and affiliations.

Bioconstitutionalism and Bioethics

A major contention of this book is that bioconstitutionalism, in contrast to traditional bioethics, allows analysts to take on board the dynamic nature of life-law relationships. All of the chapters are broadly situated in the framework of coproduction developed by scholars in STS over several decades. Coproduction refers to the joint and inseparable evolution of knowledge, materialities, and norms in contemporary societies through the fertile production of identities, institutions, discourses, and representations (Jasanoff 2004). Many examples of coproduction crowd the pages of this book: for example, the emergence of individual identities and group socialities around genetic diseases (Metzler, Reardon, Tallacchini); the formation of Embryonic Stem Cell Research Oversight committees and bodies such as Italy's Dulbecco Commission to deliberate on the ethics of research (Jasanoff, Testa); the development of new legal and ethical discourses, such as around sterilization, "postconviction DNA testing" or "group consent" to diversity research (Wellerstein, Aronson, Reardon); and struggles over how to represent, both scientifically and politically, novel borderline entities such as human-animal chimeras (Jasanoff), frozen IVF embryos (Metzler), forensic DNA databases (Winickoff), the xenotransplant patient (Tallacchini), the clinical research subject (Sunder Rajan), and even the concerned UK and European citizen (Doubleday and Wynne, Dratwa). Nowhere in these stories do we begin with firm ontologies or fixed ethical principles. Rather, each case displays a struggle for order that does justice to society's inchoate, often inarticulate moral concerns while also seeking to reap benefits from scientific inquiry and public-good technologies.

In this respect, coproduction offers an analytic purchase on law and science in the making that goes beyond traditional bioethics. Born in the evil shadows of Nazi medicine and widespread violation of research subjects' autonomy, and honed in response to new medical technologies that blur the lines between living and nonliving human bodies, classical bioethics has concerned itself mainly with protecting the taken-for-granted zones of privacy, autonomy, and integrity around the (equally taken-for-granted) individual human subject. That in itself is a task of heroic proportions, and it has perplexed some of the finest philosophical and legal minds around the world for decades (Hurlbut 2010). By representing as fluid and negotiable many of the lines that ethicists have taken as given and immutable, the essays in this volume complicate the task of bioethics in several key respects. Importantly, however, the analytic approaches espoused throughout the book also open up spaces for reflection and creative intervention by actors other than those professionally trained in normative analysis—scientists, social theorists, and lay citizens, among others—thereby contributing to the democratic governance of science and technology.

In one striking illustration of the difference between the two approaches, Alex Wellerstein's chapter shows why ethical responses to horrific practices such as compulsory sterilization should acknowledge their historical context. Sterilization does not emerge as more justifiable in Wellerstein's account of California's astonishing embrace of that supposedly therapeutic intervention. Rather, if we accept his historical explanation, responsibility is located in different and in some sense more manageable quarters: in the design of large medical institutions. Wellerstein's enthusiastic sterilizers were motivated by (to them) perfectly conscionable commitments to restoring their patients' health and social productivity; in this respect, they were acting ethically enough against the backdrop of extant cultural understandings. The problem was the centralization of institutional power in the hands of a few idiosyncratic individuals who were in a position to coproduce patient identities and therapeutic interventions without accounting to any external peers or authorities. Teaching bioethics to the California hospital administrators (in contrast to their Nazi counterparts) might not have made much difference in those historical circumstances; disciplined peer review and greater public scrutiny might have proved more effective.

Other chapters also demonstrate the contingency of ethical judgments in relation to prior institutional commitments, nested in all cases within the deeply rooted practices of living political cultures. Questions about genetic identity and its management are repeatedly framed as risk issues in the United States, and are resolved as often as not by the courts (Aronson, Wellerstein, Winickoff). After a period of laissez-faire experimentation with assisted reproduction, the Catholic Church in Italy dramatically reasserts control, both over the souls-in-waiting of IVF embryos and over the adult souls who are discouraged from going to the polls to overturn Law 40 (Metzler, Testa). National civic epistemologies determine how ethics bodies and the law deal with new biological entities such as SCNT embryos in Britain, Germany, and the United States (Jasanoff, Testa). And different balances between market and state, individual and community condition the transnational dialog on the rights and duties of xenotransplant patients in Europe and in the Anglophone worlds of Canada, Australia, and the United States (Tallacchini), as well as of pharmaceutical subjects and consumers in India and the United States (Sunder Rajan). While taking institutions as relatively stable units of analysis, all of the authors are at pains to show that large theoretical constructs, such as ideology and national political culture, rest on practices that may be deeply engrained but are always within the control of human agents. Institutions can be reimagined and rebuilt, but it takes work to do so.

Ontological Politics

Classical bioethics took the fully formed human subject as its baseline. The new intermediate entities and hybrids that modern biology creates in profusion have posed difficult challenges to that way of thinking. American policy, for instance, sought to create a specialized institutional form, the ESCRO committee, to deal with stem cells; these committees were modeled on existing bodies such as the Institutional Review Boards that supervise human subjects research and the Institutional Animal Care and Use Committees charged with ensuring the humane treatment of laboratory animals. Within a very short time, however, the ESCRO system found itself overwhelmed with new entities and issues, such as human-animal chimeras, induced pluripotent stem cells, parthenotes, and eggs donated for research. It is difficult to accommodate this rapid proliferation of new things within the analytic purview of classical bioethics, which does not take easily to ontological ambiguity for reasons outlined earlier. ESCRO committees, already uncomfortable with the extension of ethical deliberation from the relative clarity of human entitlements to the gray zone of stem cells, are still less secure in defining their institutional role with respect to novel biological entities. There is, too, a question how long such extralegal bodies can maintain their legitimacy without any public oversight; instructively, two very different European legal traditions, in Britain and Germany, have elected to place at least some aspects of bioethical analysis within bodies created by law (the Human Fertilisation and Embryology Authority and the *Deutscher Ethikrat*).

Looking at these issues through the lens of bioconstitutionalism may provide no easy or immediate answers to difficult ethical problems, but it may bring greater analytic clarity to a contested and ambiguous domain. For it can no longer be doubted that the textual revolution initiated by DNA is only now approaching its industrial potential. Techniques have come into being for reading, recombining, storing, retrieving, reprogramming, and materializing the information encoded in living things in infinitely variable ways. All this has opened up a zone of biological production between life and nonlife that law and ethics are striving to master. As the chapters demonstrate, an important function of legal proceedings has been to give names, and hence identities, to novel constructs in this ill-defined space, often by analogy to existing things and beings. Thus, Germany and Italy initially assimilated the human embryonic stem cell to the human, treating it as incipient human life entitled to protection of its integrity, whereas Britain drew the line between nonhuman and human, or object and person, only at fourteen days of embryonic development. In the United States, the xenotransplanted patient is conceived as a rights-bearer but also as a source of potentially epidemic risks, and hence in need of containment. In the more communitarian perspective of European law, the same patient is seen rather as in need of protection, through informed monitoring and surveillance by the state.

Genetic Citizens

Some of the book's most original insights relate to hitherto little remarked changes in conceptions of citizenship that have accompanied the genetic age. In this respect, the book provides important contributions to democratic theory and political thought. The authors approach the politics of biosocial transformation from many different angles, but central to all of their analyses is the demonstration that modern constitutional systems are struggling to address and accommodate new claims and questions by citizens empowered with genetic knowledge. The texts of DNA have in effect granted ordinary people new ways of reading themselves, and the accompanying ability to claim new rights vis-à-vis the state. One sees the consequence in Tessa Wick's indignant plea to the U.S. Congress ("It's like he's killing me," cited by Testa) that she be given the sovereign right to direct the production and use of stem cells derived from her own body. Comparable claims are asserted by genetically ill patients in Metzler's account of Italian stem cell politics, by convicts denied access to DNA tests in Aronson's study of recent U.S. habeas corpus decisions and in Reardon's exploration of tensions governing group consent for population-based human diversity studies.

In several of these cases, new forms of constitutional claims making are tied directly to the knowledge citizens have gained about their bodies as a result of the genetic revolution. Whether it is a patient claiming rights to unfettered biomedical research or a felon demanding access to or protection from genetic testing, it is the human body and its autonomy and integrity that have taken center stage. More subtle moves in state-citizen relations, however, are also underway as biological knowledge enables new technological interventions into life. The right to govern oneself now shifts from scenarios of hope and opportunity to scenarios of possible unintended consequences and undesirable biological control. The participatory rights demanded by British citizens in relation to GM crops, and discussed in the European Parliament and Commission in relation to the precautionary principle, showcase public disenchantment with the state's tacit social contract with science and technology. Citizens, as these and other chapters demonstrate, are no longer content simply to be told that science is advancing and will be supported by the state for public benefit. Increasingly, citizens are demanding a right to steer the very directions of scientific and technological development, asserting a right to participate in both the epistemic and the normative evaluation of competing options.

Together, these moves seem to herald a far-reaching democratization of the field of bioethics. The bioethical institutions of the twentieth century came into being within the context of a pipeline model of innovation, in which discoveries presumptively always stood for progress, and the task of ethical deliberation was only to make sure, rather late in the day, that technological applications would not unduly harm individuals. Today's participatory demands reflect a questioning of the terms of that earlier settlement. The "uninvited participation" of UK citizens that Doubleday and Wynne describe, for instance, is predicated on a loss of faith in the state's power to define the public interest without a great deal more direct input by and for the people. Public referenda from Italy to California similarly signal a desire to take the imaginaries of the genetic era away from the closed confines of expert ethics bodies, courts, and even legislative chambers into the more unruly spaces of open public debate. Bioconstitutionalism shows itself in these initiatives not as an esoteric term of academic discourse but as a dispersed and active process of reordering—indeed reconstituting—knowledge and society. The lines of contestation described in this book offer compelling insights into the conflicts that must be resolved if biopower is to be more democratically aligned with the expectations that people legitimately cherish about their bodies and selves.

References

Hurlbut, James Benjamin. 2010. *Experiments in Democracy: The Science, Politics and Ethics of Human Embryo Research in the United States, 1978–2007*. Dissertation in the History of Science. Cambridge, Mass.: Harvard University.

Jasanoff, Sheila, ed. 2004. *States of Knowledge: The Co-Production of Science and Social Order*. London: Routledge.

Basic Bioethics

Glenn McGee and Arthur Caplan, editors

Books Acquired under the Editorship of Glenn McGee and Arthur Caplan

Peter A. Ubel, *Pricing Life: Why It's Time for Health Care Rationing*

Mark G. Kuczewski and Ronald Polansky, eds., *Bioethics: Ancient Themes in Contemporary Issues*

Suzanne Holland, Karen Lebacqz, and Laurie Zoloth, eds., *The Human Embryonic Stem Cell Debate: Science, Ethics, and Public Policy*

Gita Sen, Asha George, and Piroska Östlin, eds., *Engendering International Health: The Challenge of Equity*

Carolyn McLeod, *Self-Trust and Reproductive Autonomy*

Lenny Moss, *What Genes Can't Do*

Jonathan D. Moreno, ed., *In the Wake of Terror: Medicine and Morality in a Time of Crisis*

Glenn McGee, ed., *Pragmatic Bioethics, 2nd edition*

Timothy F. Murphy, *Case Studies in Biomedical Research Ethics*

Mark A. Rothstein, ed., *Genetics and Life Insurance: Medical Underwriting and Social Policy*

Kenneth A. Richman, *Ethics and the Metaphysics of Medicine: Reflections on Health and Beneficence*

David Lazer, ed., *DNA and the Criminal Justice System: The Technology of Justice*

Harold W. Baillie and Timothy K. Casey, eds., *Is Human Nature Obsolete? Genetics, Bioengineering, and the Future of the Human Condition*

Robert H. Blank and Janna C. Merrick, eds., *End-of-Life Decision Making: A Cross-National Study*

Norman L. Cantor, *Making Medical Decisions for the Profoundly Mentally Disabled*

Margrit Shildrick and Roxanne Mykitiuk, eds., *Ethics of the Body: Post-Conventional Challenges*

Alfred I. Tauber, *Patient Autonomy and the Ethics of Responsibility*

David H. Brendel, *Healing Psychiatry: Bridging the Science/Humanism Divide*

Jonathan Baron, *Against Bioethics*

Michael L. Gross, *Bioethics and Armed Conflict: Moral Dilemmas of Medicine and War*

Karen F. Greif and Jon F. Merz, *Current Controversies in the Biological Sciences: Case Studies of Policy Challenges from New Technologies*

Deborah Blizzard, *Looking Within: A Sociocultural Examination of Fetoscopy*

Ronald Cole-Turner, ed., *Design and Destiny: Jewish and Christian Perspectives on Human Germline Modification*

Holly Fernandez Lynch, *Conflicts of Conscience in Health Care: An Institutional Compromise*

Mark A. Bedau and Emily C. Parke, eds., *The Ethics of Protocells: Moral and Social Implications of Creating Life in the Laboratory*

Jonathan D. Moreno and Sam Berger, eds., *Progress in Bioethics: Science, Policy, and Politics*

Eric Racine, *Pragmatic Neuroethics: Improving Understanding and Treatment of the Mind-Brain*

Martha J. Farah, ed., *Neuroethics: An Introduction with Readings*

Books Acquired under the Editorship of Arthur Caplan

Sheila Jasanoff, ed., *Reframing Rights: Bioconstitutionalism in the Genetic Age*

Contributors

Jay D. Aronson is Associate Professor of Science, Technology, and Society at Carnegie Mellon University. His research and teaching focus on the interactions of science, technology, law, and human rights in a variety of contexts. His first book, *Genetic Witness: Science, Law, and Controversy in the Making of DNA Profiling* (Rutgers University Press, 2007), examines the development of forensic DNA analysis in the American legal system. He is currently engaged in a long-term study of the ethical, political, and social dimensions of postconflict and postdisaster DNA identification of the missing and disappeared. He received his PhD in History of Science and Technology from the University of Minnesota and was a postdoctoral fellow at Harvard University's John F. Kennedy School of Government.

Robert Doubleday is Senior Research Associate in the Department of Geography at the University of Cambridge. His recent research, funded by the Wellcome Trust, explores how academic scientists working on nanotechnologies understand and respond to the political and policy contexts in which they work. Previously he spent a year as a policy fellow at the UK Government Office for Science; two years as a "lab-based social scientist" at the University of Cambridge Nanoscience Centre; and one year on a Fulbright Scholarship at Harvard University's John F. Kennedy School of Government.

Jim Dratwa's research and publications address issues of transnational expertise, legitimacy, and governance, probing the interfaces between science, citizenship, and policy making through ethnographic inquiries into international organizations. For the last ten years, he has combined academic and policymaking activities. He has taught at the École des Mines de Paris, Harvard University, the Université Libre de Bruxelles, and the Facultés universitaires Saint-Louis, Brussels, where he is currently Professor of Political Science. He is also a civil servant with the European Commission, at the department for research policy, in the team working on policy analysis and prospective analysis.

Sheila Jasanoff is Pforzheimer Professor of Science and Technology Studies (STS) at Harvard University's John F. Kennedy School of Government. She has authored more than 100 articles and book chapters and is author or editor of a dozen books, including *The Fifth Branch* (Harvard University Press, 1990), *Science at the Bar* (Harvard University Press, 1995), and *Designs on Nature* (Princeton University Press, 2005). Her work uses innovative methods of cross-national comparison and legal analysis to illuminate the role of science and technology in constructing norms of evidence, persuasion, and public reason in modern democracies. She was founding chair of the STS Department at Cornell University and has held distinguished visiting appointments at numerous American and European academic institutions. She holds AB, PhD, and JD degrees from Harvard.

Ingrid Metzler is a researcher at the Life-Science-Governance Research Platform at the University of Vienna. She has studied political science at the University of Rome, *La Sapienza*, the University of Innsbruck, and the University of Vienna. In 2005, she was a Marie Curie fellow at the Science and Technology Studies Unit at the Department of Sociology of the University of York. Her research focuses on the framing of issues at the intersection of science, technology, and politics.

Jenny Reardon is Associate Professor of Sociology and Faculty Affiliate in the Center for Biomolecular Sciences at the University of California, Santa Cruz. Her research investigates how novel forms of technoscience (such as genomics and nanotechnology) are constituted along with novel forms of governance and modes of constructing human identity. She is a primary organizer of the Science and Justice Working Group at the University of California, Santa Cruz. Her first book, *Race to the Finish: Identity and Governance in an Age of Genomics*, was published with Princeton University Press in 2005. She is currently working on a second book manuscript entitled *The Post-Genomic Condition: Technoscience at the Limits of Liberal Democratic Imaginaries*.

Kaushik Sunder Rajan is Associate Professor of Anthropology at the University of Chicago. He is the author of *Biocapital: The Constitution of Postgenomic Life* (Duke University Press, 2006) and the editor of *Lively Capital: Biotechnologies, Ethics and Governance in Global Markets* (Duke University Press, 2011). His work explores the relationships between the life sciences and global capital, with a specific empirical focus on the United States and India. He is currently working on a number of projects relating to aspects of pharmaceutical development in the Indian context, such as global clinical trials, intellectual property regimes, and translational research.

Mariachiara Tallacchini is Professor of Legal Philosophy at the Law Faculty of the Università Cattolica S.C. of Piacenza and of Bioethics at the Faculty of Biotechnology of the Università degli studi of Milan (Italy). Her interests concern the legal regulation of science and technology in the domain of life sciences. She chaired the Committee on Science in Society for the European Commission's FP7 and is a member of the Ethics Committee of IXA (International Xenotransplantation Association). She has coauthored or coedited several books, including *Ethics and Genetics* (Berghahn 2003), *Biotecnologie. Aspetti etici, sociali, ambientali* (Mondadori 2004), and *Trattato di Biodiritto* (Giuffrè 2010).

Giuseppe Testa holds an MD and a PhD in Molecular Biology and an MA in Bioethics and Law, and has been visiting fellow at Harvard University's John F. Kennedy School of Government and Wissenschaftskolleg Berlin. He heads the Laboratory of Stem Cell Epigenetics at the European Institute of Oncology (Milan), where he cofounded an interdisciplinary PhD program on Life Sciences, Bioethics, and Society. His STS and bioethics scholarship focuses on the relationship between the life sciences and the evolution of modern democracies. His scientific and bioethics/STS work has appeared in leading peer-reviewed journals. He is the author, with Helga Nowotny, of *Naked Genes: Reinventing the Human in the Molecular Age* (MIT Press, 2011).

Alex Wellerstein is a postdoctoral fellow at the Managing the Atom Project, Belfer Center for Science and International Affairs, at Harvard University's John F. Kennedy School of Government. He primarily works on the history of nuclear weapons, scientific secrecy, and relations between science and the state. He received a PhD in the History of Science from Harvard University in November 2010, with a dissertation entitled "Knowledge and the Bomb: Nuclear Secrecy in the United States, 1939–2008" and a BA in History from University of California, Berkeley, in December 2002.

David E. Winickoff is Associate Professor of Bioethics and Society at the University of California, Berkeley, where he codirects the Science, Technology, and Society Center. Trained at Yale, Harvard Law School, and Cambridge University, he has published over thirty articles in leading bioethics, biomedical, legal, and science studies journals such as *The New England Journal of Medicine*, the *Yale Journal of International Law*, and *Science, Technology, and Human Values*. His academic and policy work spans topics of biotechnology law, intellectual property, the World Trade Organization, food safety regimes, and human subjects research.

Brian Wynne is Professor of Science Studies and Associate Director of the UK ESRC Centre for Economic and Social Aspects of Genomics at Lancaster University. His work involves sociological analysis of scientific knowledge in public arenas, such as environmental policy and controversy, risk assessment and regulation, and sociological interpretation of public attitudes and responses to science-intensive innovations and policies. He has published and edited several books on scientific knowledge in public policy settings, including *Rationality and Ritual: Participation and Exclusion in Nuclear Decision Making* (1982). He has authored approximately fifty peer-reviewed articles and been actively engaged in UK and European science and environmental policy.

Index